PENGUIN BOOKS

THE STORM

Ivor van Heerden was born in Johannesburg, South Africa. He is cofounder and deputy director of the Louisiana State University Hurricane Center and director of the Center for the Study of Public Health Impacts of Hurricanes. He is also associate professor of civil and environmental engineering at LSU. He holds a Ph.D. in marine sciences from LSU, where his research focused on the Atchafalaya River Delta; his ongoing research areas include disaster preparation and response, coastal geomorphology, environmental management, and habitat restoration. He lives near Baton Rouge, Louisiana.

Mike Bryan has written or collaborated on many books, including Cal Ripken's bestselling autobiography, *The Only Way I Know*; *Uneasy Rider*; and *The Afterword*, a novel.

THE STORM

What Went Wrong and Why During Hurricane Katrina—the Inside Story from One Louisiana Scientist

IVOR VAN HEERDEN

and

MIKE BRYAN

With Field Sketches
by the Author

PENGUIN BOOKS

PENGUIN BOOKS
Published by the Penguin Group
Penguin Group (USA) Inc., 375 Hudson Street, New York, New York 10014, U.S.A.
Penguin Group (Canada), 90 Eglinton Avenue East, Suite 700, Toronto,
Ontario, Canada M4P 2Y3 (a division of Pearson Penguin Canada Inc.)
Penguin Books Ltd, 80 Strand, London WC2R 0RL, England
Penguin Ireland, 25 St Stephen's Green, Dublin 2, Ireland (a division of Penguin Books Ltd)
Penguin Group (Australia), 250 Camberwell Road, Camberwell,
Victoria 3124, Australia (a division of Pearson Australia Group Pty Ltd)
Penguin Books India Pvt Ltd, 11 Community Centre, Panchsheel Park, New Delhi – 110 017, India
Penguin Group (NZ), 67 Apollo Drive, Rosedale, North Shore 0745, Auckland,
New Zealand (a division of Pearson New Zealand Ltd)
Penguin Books (South Africa) (Pty) Ltd, 24 Sturdee Avenue, Rosebank, Johannesburg 2196, South Africa

Penguin Books Ltd, Registered Offices:
80 Strand, London WC2R 0RL, England

First published in the United States of America by Viking Penguin,
a member of Penguin Group (USA) Inc. 2006
This edition with a new afterword published in Penguin Books 2007

1 3 5 7 9 10 8 6 4 2

Maps on pages 70–71, 74–75, and 270–71 by Adrian Kitzinger

ISBN 0-670-03781-8 (hc.)
ISBN 978-0-14-311213-6 (pbk.)
CIP data available

Printed in the United States of America
Designed by Nancy Resnick

This book is dedicated to those who lost their lives during Hurricane Katrina and to their families.

The book is also for the first responders, who unselfishly did their best against the odds to save lives.

I also dedicate this book to disaster science researchers everywhere, who follow their passion even under duress and never lose sight of the ball.

And finally, I dedicate this book to my mother, Ivy Phyllis Bebb, who taught me to always stand by one's principles. She also taught me the power of prayer.

CONTENTS

Religion that God our Father accepts as
pure and faultless is this:
To look after orphans and widows in their distress
And to keep oneself from being polluted by the world.

James 1:27

INTRODUCTION

DISASTER, TRAGEDY, FAILURE—AND HOPE

By eight o'clock Monday night, August 29—almost fourteen hours after the landfall of Hurricane Katrina—even I was tempted to join in the back slapping at the state's Emergency Operations Center (EOC) in Baton Rouge. Using every available megaphone, I'd been warning for years about the inevitable catastrophe that would befall New Orleans and southeastern Louisiana: a total drowning. So had all of my colleagues and many other scientists who had studied the lay of the land. It was bound to happen, sooner or later. It could have happened with Katrina, if she had tracked just twenty miles to the west and on a northwesterly course. Earlier studies using our now famous storm-surge computer model at LSU had showed that hypothetical catastrophe clearly. On Katrina's actual course, our model still predicted the flooding in New Orleans and in the parishes to the east and south, but most of these areas had flooded before, never disastrously. They could be drained quickly, with minimal permanent damage. On Monday night, this is what we thought.

Communications were suspect. After the National Weather Service office in Slidell, across Lake Pontchartrain from New Orleans, issued its warning at 8:14 A.M. Monday morning of a breach in the levee along the Industrial Canal, that office lost power. The office in Mobile, Alabama, then took over, but apparently the warning got lost. At the EOC, we knew that the Mississippi coastline to the east would have been essentially wiped out, but here's the blunt truth: Our attention was focused on New Orleans, and not simply because we lived and worked in Louisiana. Most—I would hope *all*—of the professionals in the center were aware of New Orleans's particular peril, and some had devoted a fair portion of their careers to studying the city's vulnerability, partly because it is a fascinating subject with life-and-death consequences. And now the Crescent City had apparently managed to keep its head above water once again; it would live to sweat out the next big storm. So it seemed at eight o' clock Monday night, and I was packing up to leave when a young staffer walked into our cubicle at the EOC and said he'd just picked up a call from a nursing home that had taken in two feet of water, and it had risen half a foot in just the last hour.

Fresh or saline? That was my first thought, because it's always the question about unwanted water in New Orleans. If it's fresh, it's rainwater—a normal flood; if salty, Lake Pontchartrain (which is actually brackish), and this might mean a serious breach of the lakefront levee system on the northern side of the city, or a breach from Lake Borgne to the east. We didn't have the results of the taste test in the nursing home, but surely the Army Corps of Engineers, which had built the levees and whose cubicle in the EOC was right next to ours, would have known about a lakeside breach and somehow been able to spread the word. On the other hand, Katrina and the heavy rain were long gone. Why the rapidly rising water *now*? The tiniest little chill ran up my spine. But I had nothing to pin it on and no good way to find out much more, so at 9:30 P.M. I hit the road to join my family at our home twenty miles east of Baton Rouge. I had no idea what had happened out there in New Orleans.

But then, I *did* know. Driving home. Thinking again about that suddenly rising water. How could rainfall runoff possibly cause that flooding? It couldn't. There must have been a new breach in a levee. But surely the Corps would know about it! Surely their people and the local levee boards were monitoring every foot of the 350 miles of levees that protected the city from the Mississippi River and Lake Pontchartrain. But then, why had the Corps's staffers said nothing?

When the worst had been expected in New Orleans over the weekend, numerous officials had warned people who weren't leaving town to be sure they had an ax handy, because they were going to need it to chop their way out of the attic. They were trying to scare folks. Now their warning scared me. If there had been a major levee breach anywhere, thousands and thousands of people were going to bed in the dark, thinking the worst was over, and they would wake up in the middle of the night to a horrible discovery and be forced into those attics.

At my home, all was relatively secure—a few trees down, one close call, but no roof damage. With no electricity, land phone, cable, or Internet, with only spotty cell coverage, the isolation was almost complete—and welcome. I'd finally be able to sleep. But I couldn't sleep. I was thinking about those people scrambling out of the rising water—if they could scramble. The call had come from a nursing home. How were those folks going to *scramble* anywhere? Did the able-bodied have axes? Could they wield them? And would the attic be high enough? Would the *roof* be high enough? In the Lower Ninth Ward and parts of St. Bernard Parish, maybe not. Lying in bed, I envisioned the deaths of thousands.

I did finally fall asleep, and it was almost noon when I got in the trusty Xterra to drive back to the EOC. The sun was shining, the air was calm. Maybe I was wrong. Maybe the nursing home report was some kind of fluke. I turned on the radio for the latest—and my heart sank for good. Something terrible *had* happened with the levees on Monday, and no one had told us. Water was pouring into the city. Should I have turned around the night before and returned to the EOC? I'll ask myself that question for the rest of my life and reconcile my failure to do so with the knowledge that Monday night was much too late to spread the alarm—because there was no way to spread the alarm. Battery-powered radios could pick up a few channels, including locally famous WWL, but otherwise the city had been in the dark. With no electricity, everyone would have gone to bed.

As I drove to Baton Rouge, I began getting angry. As the days advanced, I got angrier. New Orleans had not even been the bull's-eye for this storm, which also had turned out to be less powerful than expected. Nevertheless, much of the city was going under, with the whole world watching in disbelief. How could the United States of America have left one of its crown jewel cities so vulnerable to a *pre-*

ventable disaster that I and many others had been warning about for years? How could this nation have been so unprepared for the aftermath? Hurricane Katrina was both a natural disaster and a systemic failure on the part of our society. Together, they produced tragedy.

This book is about both the storm and the failure—and the tragedy. I am a disaster science specialist and hurricane researcher who tends to wear his heart on his sleeve. I rarely hold my tongue. I rarely see any good reason to, and certainly not in this case. As a scientist, I champion a reality-based view of the world, old-fashioned as that may be, and marshy, swampy coastal Louisiana is the very definition of an inherently vulnerable landscape. It has always been susceptible to the ravages of hurricane winds, storm surges, and the invasive activities of a certain species of mammal (and I don't mean nutria). Large neighborhoods in New Orleans had flooded during Hurricane Betsy in 1965, and eighty-one folks had drowned. But then the Army Corps of Engineers beefed up the levees and there had been no other major flooding from storm surges in forty years—just the fairly regular flooding from the torrential rain the city expects several times a year. People had let down their guard. After all, New Orleans is—or was—the city that care had forgotten anyway. On the other hand, some of the local geographers and oceanographers and engineers and the like—you know, the pocket-protector crowd, as we were known before computers—started to suspect that Louisiana was not only more vulnerable to devastation than its citizens wanted to admit, but far more vulnerable *than it had been,* because the wetlands that buffer the inland zones, including New Orleans, were disappearing at an alarming rate. Restoring these coastal wetlands was one key to long-term alleviation of surge flooding, but could the state pull off such a complex endeavor requiring billions of dollars and the total commitment of the citizens.

In 1994, when I took over as head of the state's coastal restoration program, I decided to find out. My abiding faith was simple (and it still is): The catalyst to compromise is a thorough understanding of the science. Within eighteen months, however, I was gone and our new, comprehensive initiative was dead—details forthcoming, but which I can summarize here as the petty politics of junior bureaucrats who had the ear of the official decision makers and of politicians bearing grudges. Our plans for the vital coastal restoration fell apart, but we had to keep trying.

In 1998, Hurricane Mitch hit Honduras with the torrential rains and flooding that killed more than ten thousand people, left one million homeless, and caused massive damage to that country's fragile infrastructure. As it happens, then-president Carlos Flores is an alumnus of LSU; his wife a native of Louisiana. Prompted by some Honduran expatriates and Bruce Sharkey of LSU's landscape architecture department, Dr. Lynn Jelinski, the university's vice chancellor for research at the time, asked an ad hoc group of researchers to fly down to assess the situation. I was on that team, working for the Louisiana Geological Survey. Also on the trip was Marc Levitan, professor of civil engineering, premier "dynamic wind" expert, and a complete prince of a fellow, and Emir Macari, then head of the civil and environmental engineering department. The Hondurans had great ideas but lacked the governmental support structure to make things happen. We found many ways that LSU could help, but we also found out that the "Beltway Bandits" with the good connections to the U.S. Agency for International Development were going to get all the work. Once again, a lot of American aid would mainly help certain corporations get on their feet, rather than address the real issues in Honduras, where the poorest people live in the most flood-prone areas.

Marc Levitan and I had never met before this trip, but it turned out that we had a lot in common besides hurricanes. We are both turned on by applied research, stuff that's applicable to real-world problems. We both enjoy working with local and state governments as well as the chase for funding from some of the more "serious" sources. It also turned out that each of us had recently approached Lynn Jelinski with different but related ideas, Marc for a center to research hurricanes, me for one for applied sciences that would cover hurricanes, coastal restoration, and environmental issues. Over beers one night in Honduras, we decided to join forces. Jelinski represented a wonderful window of opportunity in the upper administration at LSU. Very early in her tenure, Jelinski, who had come to Louisiana from Cornell, could tell that the state had many challenging problems. She had great vision, and she understood the potential of applied science—an exception to the rule on campus. (In its quixotic attempt to become the Harvard of the Bayous, LSU does not give much respect to "applied" research. And there is almost a disdain for working with and for local agencies and governments.)

One year later the LSU Hurricane Center was a reality, with Marc the director and myself the deputy director. Marc has always stood by me when I've had my tiffs with LSU. He has never lost sight of the goals we set for our work. During every emergency he has always been a barrel of energy and an inspiration.

Shortly before the original hurricane center got rolling, Louisiana's Board of Regents was given tobacco settlement monies to set up a public health research trust fund. Along with many other suitors, of course, I and a multidisciplinary team of scientists from three different universities applied for funding to set up a sister institution to the Hurricane Center, one that would focus on public health issues related to the big storms. We were passed over in the first round because, we were told, our scope was too broad. I refined our proposal, focusing just on New Orleans, and the following year we received $3.65 million for five years. The Center for the Study of Public Health Impact, commonly known as the Hurricane Public Health Center, opened in 2002.

Each center is a "virtual" organization. We don't have a building. We don't even have a coherent suite of offices. We have only a group of dedicated scientists from many fields working to understand all aspects of a major hurricane strike on southeast Louisiana. What, exactly, would be the effects? What could we do to prepare for and mitigate them? For all of the alarms raised over the preceding years about the dire vulnerability of New Orleans, no group had analyzed the threat in all its complexity. This became our job. We are now known for the computer models that predict with astonishing accuracy the storm surge that can be expected from a hurricane of a given strength approaching any section of the Gulf Coast, but we have also studied problems with evacuation, public information, contamination (air, water, and soil), housing, stray animals, infectious diseases, the famous levees, and much more. We set out to demonstrate the incredible challenge of both preparing for the most dangerous storms and dealing with their aftermath. We always believed that both tasks would be exponentially more difficult than the emergency management establishment seemed to grasp. Our research supported this belief. Then Katrina proved the point.

As our work continued and our research was published and disseminated by all possible means, some people and organizations were willing to listen, some weren't. Among the latter was FEMA, the Federal Emergency Management Agency. In July 2004, the agency paid in excess of five hundred thousand dollars for an exercise in which numerous agencies at all levels of government worked for eight days on their response to a hypothetical hurricane that had catastrophically flooded New Orleans. The sorry story of "Hurricane Pam" has received some press coverage already, and it will receive more—in this book, if nowhere else. As much as we at LSU tried to get the latest and best science into the picture during this exercise, we were never really given an opening. I, specifically, was the foreign geek, the guy from South Africa with the odd first name and the odd last name and the odd accent. Such was my impression, fixed forever by the woman from FEMA I was trying to convince of the need to plan for short-term housing for hundreds of thousands of evacuees—tent cities, I called them. She sarcastically answered, "Americans don't live in tents." Okay, then, *McMansions*, but these people are going to need a roof over their heads! And so they did.

Immediately after the storm, I received permission to shift some of our funding from research about catastrophic hurricanes to an operational response to this particular one. Our new job was to provide services and expertise for the recovery effort. Our Geographic Information Systems (GIS) database—seventy categories of information, any combination of which can be layered on a map—was used by FEMA and other agencies as they scrambled to accomplish what they should have been well prepared to accomplish years earlier.

Landfall for Katrina was Monday morning. As the flooding in New Orleans spread on Tuesday and the pathetically inadequate response became more evident by the hour, my mood lowered. Along with some of my colleagues, I wanted to scream in frustration and, yes, in a bit of self-vindication, "We told you so!" On one of my first tours of the submerged neighborhoods I waded past a house in the Lower Ninth Ward in which the only possessions still above water level were the family photographs on a high mantelpiece. The water was fetid, the air was rancid, I had seen a floating body not a block away, and there, right in the middle of this apocalyptic disaster, was a surreal vision through the open window of gowned graduates, smiling brides and grooms, proud parents and grandparents, happy babies. For some reason, this was the scene in the stricken city that put me over the edge and broke my heart. Where were these people now? If even alive, what future did they have? How had they been served by their government? I felt I could and should speak for them, and I did.

As the director of the Hurricane Public Health Center and, more to the point, as one of the more notorious Cassandras of recent years, I was a pretty obvious target for the hundreds of reporters soon on the scene—print, radio, television, local, regional, national, international—who were also asking these questions on behalf of their audiences. In retrospect, I think I became a popular interview because I was (and remain) a straight-ahead guy who calls it the way I see it, and in this instance my call coincided with the conclusions the reporters were rapidly drawing for themselves. As the "natural disaster" story evolved into the "national disgrace" story, the reporters lost patience with the politicians and bureaucrats who spent the first five minutes of each press conference thanking all the other politicians and bureaucrats arrayed left and right for their great work. The reporters welcomed me as someone willing to state bluntly that some of those officials had failed in their responsibilities, that assorted government agencies had ignored years of excellent science, failed to heed warning after warning, failed to plan for the disaster, failed to act when it did happen, and, if the past is indeed prologue, would probably now fail to rebuild New Orleans properly and assure its safety from another catastrophe.

Suddenly I found myself on television all the time. On *Larry King Live* the first Friday night after the storm—my third night in a row on the show—I answered Larry's question about the failed response so far and then said, "We in Louisiana can only trust that our governor, and especially the president, are putting all the resources of the federal government and our *mighty military* to bear on this problem." I laid on the sarcasm in that final phrase, and, naturally, I like to believe that the arrival of the first large contingent of military helicopters the following morning was not coincidental. When I watched the scenes of white cops confronting at gunpoint groups of mostly black residents, I flashed back to startlingly similar scenes of apartheid South Africa, where I grew up and went to school. Don't get me wrong. I'm not drawing a straight analogy between apartheid in the 1980s and racial attitudes in my adopted state of Louisiana in 2005. I say only that some of the ugly scenes in New Orleans had undeniably racial overtones, and I got angry, and the anger helped me find fresh energy to press ahead and continue communicating the whole story.

Subsequently, as producers and reporters started to investigate the specific causes of the levee breaches that were responsible for *87 percent* of the flooding, by volume, in New Orleans proper, I didn't hesitate to point their efforts in what I was pretty sure was the right direction—that is, *not* the direction in which the Army Corps of Engineers was trying to lead everyone. Various Corps officials insisted for weeks that the storm surge from Katrina had overtopped and overwhelmed their Grade-A levees. Our team thought the levees had failed, plain and simple— *buckled*—for reasons we had to determine. Over the following months five groups conducted research into the issue: the University of California—Berkeley team funded by the National Science Foundation; the American Society of Civil Engineers (ASCE); the Senate Homeland Security and Governmental Affairs Committee; the Corps; and the state of Louisiana (led by our team at the LSU Hurricane Center). A consensus developed. Our team was correct. The truth did get out. I honestly didn't care whether making sure that it did cost me my job, and for a while it looked as if it might.

Over thirteen hundred citizens in Louisiana and Mississippi died due to Hurricane Katrina—the number as of February 2006, and certain to go up, perhaps dramatically, as the missing are reclassified. Six months after the storm, one hundred thousand families were still homeless. Some of those deaths and some of those dislocations were inevitable, because Katrina was a natural disaster. Others—the majority—were man-made. I don't see how we can avoid that conclusion. The levee systems failed inexcusably. We now thoroughly understand the need for coastal restoration as a buffer against the big storms, but land loss continues at an alarming rate. So what next? Should we clean up and rebuild New Orleans—to the extent even possible—if we then repeat the mistakes of the past? No, because the point overlooked in much of the Katrina media coverage is the fact that this hurricane was *not* the big one. I've learned that people don't want to hear this, because it makes them angry. But there it is. As Katrina *should have* affected New Orleans proper, she was decidedly a medium hurricane. Sometime in the foreseeable future a bigger storm will not take that last-minute jog to the east and every square foot of New Orleans—all of it, not just 80 percent—will be underwater, and deeper underwater than this time. Unless, that is, the right measures are authorized and funded immediately, then executed promptly and properly.

I don't like to see good science pushed to the sidelines just because it conflicts with narrow interests pushing their self-serving agendas. Such politics as usual helped to inundate New Orleans in 2005. If science and engineering had been allowed to play their proper role in the development of policies for the wetlands and the levees, we wouldn't be in this situation today. If nothing changes in the future, one fifth of the state of Louisiana—everything south of Interstate 10, including the city of New Orleans in its entirety—will disappear beneath the waves, gone for good, and we will have no one to blame but ourselves. Future historians will be writing books about the "Cajun Atlantis."

Hard as it is to believe, nature has actually given us a bit of a second chance. There's something left to work with in New Orleans. We must put aside the politics, egos, turf wars, and profit agendas if we're going to reconstruct this city effectively, engineer proper levees, and restore the buffering coastline.

This wedge of the continent has been changed forever, physically, economically, culturally. What will the new city be like? Will it become Nuevo Orleans (undocumented Mexican laborers are already the dominant cleanup force throughout the city, living ten to a room, conceivably to stay around and replace blacks as the dominant "minority"), or Six Flags Over New Orleans, or something else? And the former residents who can't afford a ticket to the new city, or just don't want one—where do they end up, and in what circumstances? These questions I can't answer. From my vantage pretty near the center of the story, I'll stick to those I can.

I am not a reporter. In this context, I am not even objective. What follows is not a diary or a recapitulation of events during the Katrina emergency. It is my assessment of how we got to this truly tragic moment—the story I have come to think of as the nearly perfect folly—and how we must proceed from here.

ONE
STORM CLOUDS

Despite their incredible power—Katrina generated energy equivalent to one hundred thousand atomic bombs on her journey across the Gulf of Mexico—hurricanes are actually quite fragile. For that matter, so are the bombs. The design and machining tolerances required to produce a blast instead of a squib are minuscule. Both the storm and the bomb are therefore living proof of the main tenet of chaos theory: Small changes in a system may have large consequences down the line. (Did a butterfly just flutter past my window?) We expect about ninety or so tropical storms to develop worldwide every year. In 2005, the Atlantic Basin alone produced a record twenty-seven tropical storms, with a record fifteen developing into hurricanes. The public asks why so many. Researchers wonder why so few.

Each system begins as a thunderstorm or group of thunderstorms, which are more or less ubiquitous in the tropical summer and common enough in what passes for winter. The reason is simple: The warm air rises, and as it rises it cools, and as it cools it can hold less water vapor, which therefore condenses into droplets of water, which then fall as rain. The condensation releases latent energy in the form of heat, which then reinforces the dynamics of the storm. The visible evidence is the cumulus and cumulonimbus clouds that presage or confirm the presence of thunderstorms. They are a daily affair—but then they're gone, because the storm has a built-in limiting factor: Downdrafts of cool, dry air from the heights block the system from drawing in enough warm, moist air from the surface to sustain the action. Thus the "heat engine" utilizes most of the available resources of heat and humidity within half an hour, perhaps an hour, and quickly fades into unwritten history, with the tourists immediately returning to the beach or the homeowner to the porch to enjoy the sunset. All of us who live along the Gulf or Atlantic coasts can loll away many a pleasant summer afternoon watching this physics lesson unfold (but not those who live on the West Coast, where summer is the dry season, because this region is dominated by cool maritime polar air masses that suppress the necessary convectional uplift over land).

In the late summer of 1989 I also lolled away some of those afternoons in the U.S. Virgin Islands, where I was studying the suffering coral reefs of the Buck Island National Reef Monument off the northeast coast of St. Croix, an environment about as tropical as tropical can get, as I learned on the very day I sailed into the port at Christenstead after the six thousand–mile voyage from South Africa. I'd left home with my sailing partner, Nan, six months before, and that's exactly how long I had—six months—to get the official U.S. port of entry stamp that would activate my green card, that wonderful document that is the initial step to acquiring U.S. citizenship. So the whole trip west across the South Atlantic and then north across the equator was a slow race against time (remember, a sailboat averages maybe 8 mph to 10 mph; a bicycle is faster—except on water). On September 19, 1989, a few days before we arrived in Christenstead and a few more days before I had to obtain my documents, St. Croix was hit hard by Hurricane Hugo, a Cat 4 storm (lingo for Category 4 on the famous Saffir-Simpson scale, on which Cat 5 is the strongest storm). I had no idea. South Africans rarely experience cyclones, as tropical storms that form over the Indian Ocean are called. (The exception was Cyclone Demoina, which struck the Zululand coast in October 1984. The winds were minimal but the rains were not; forty inches in twenty-four hours cost four hundred people their lives.) At Christenstead on St. Croix, what an unbelievable sight as we sailed in, with boats tossed everywhere and buildings destroyed. The *Maggie* seemed to be the only sailboat afloat in the whole harbor. Welcome to the United States!

Destruction or no destruction, I had my paperwork problem and had to find the immigration lady. I eventually did, and she was very helpful after I dragged a couple of chairs from the wrecked office and set them up on the wharf on a bright sunny day—as they usually are in the wake of a storm. The necessary forms were water-stained, but they worked. I was home free. The next day I took up temporary residence at the St. Croix Yacht Club, or what was left of it, which was very little besides lots and lots of stunned snowbirds. Soon I was living aboard the *Maggie*, a couple of hundred yards off the pier of the West Indies Marine Lab. I dived all day with some great folks, did good work, I believe, sipped Cruzan rum cocktails in the evenings, and shared potluck dinners with other cruisers. I had loved growing up in South Africa, but this life in the tropics was an unexpected dream come true. Still, all of us always had an eye on the weather and had checked out the available hurricane holes in which to stash our boats.

In order for a group of thunderstorms to organize into an official low-pressure zone—a "tropical depression"—some trigger must serve to alleviate the drying influence of the downdrafts. One such trigger can be an easterly wave floating westward across the Atlantic from the hot waters off the African coast. As many as one hundred of these can form every season, each a freestanding aberration in barometric pressure, a disturbance that can induce a group of thunderstorms to, in effect, wrap around themselves and thereby get something bigger going. The odds that any given cluster of thunderstorms will get organized are very low, but with two thousand clusters breaking out around the world every single day, why only ninety tropical storms in an entire year worldwide? As I said, these things are fragile.

The easterly wave floats along . . . still floating . . . one thousand miles . . . nothing happening, smooth sailing . . . but then, when conditions are just right—when the chaos is just right—this wave or maybe the tip end of a low-pressure trough or some other mechanism we don't fully understand (there are numerous candidates) gently pulls the trigger on the rapidly rising column of air at the center of the disturbance, which has a lower pressure than the surrounding air, and this low-pressure action sucks in more air, and this rapidly moving air evaporates seawater from the ocean, bringing still more moisture and heat into the system—the heat as a result of friction, both from the action of the wind on the water and from the molecules in the atmosphere rubbing against each other—and the planet below spinning toward the east induces a cyclonic, counter-clockwise rotation to the whole thing.

That's when the National Hurricane Center (NHC) in Miami takes official notice, and thus was a certain disturbance over the eastern Bahamas christened as the twelfth tropical depression of the 2005 Atlantic season, on Tuesday, August 23, 2005. Advisory #1 pin-pointed the center near latitude 23.2 north, longitude 75.5 west—about 175 miles southeast of Nassau in the Bahamas. The depression was moving toward the northwest, with some strengthening likely. All interests in the Bahamas and southern Florida were advised to pay close attention, because further watches and warnings would probably be posted soon.

The NHC forecasters had very little doubt about the general direction this new depression would take. What a change from one hundred years ago—1900, specifically, when a hurricane with no name (these were not utilized until 1950) entered the Gulf of Mexico headed toward the northwest. The renowned Cuban hurricane researchers, whose tradition of excellence had been fostered in the local Jesuit community, had tracked this hurricane as it stormed along the length of their island and had predicted its track into the Gulf, but the American forecasters insisted it was 150 miles northeast of Key West and threatening the Atlantic Seaboard. Four days after these forecasters warned fishermen in New Jersey to stay in port, the storm struck Galveston 1,500 miles to the southwest, flattening that Texas beach resort and port, killing at least eight thousand people—the deadliest storm tragedy in U.S. history, and the subject of *Isaac's Storm*, by Eric Larsen, the most popular hurricane book ever published.

We can fault the American forecasters for their xenophobia about the Cuban team (even before Castro), but otherwise their ignorance just reflected the state of the art at the time. If a storm didn't leave a trail of firsthand reports as well as actual destruction while crossing the islands of the Caribbean or the Bahamas, forecasters had no way of even knowing where it was, or even *if* it was. This shortage of reliable information improved dramatically in 1909, with the introduction of the maritime radio, which allowed ship captains to alert forecasters to deteriorating local conditions. Given enough of these reports, forecasters might get a good idea about present location but still have no clue about destination. In 1944, the first reconnaissance aircraft flew into storms to take measurements, and at about the same time, radar, newly invented by the British in World War II, was able to pick up storms as they approached the coastline. In the 1960s, satellite images revealed storms in remote areas of the tropics, storms whose very existence might have escaped detection otherwise, and by the midseventies the geostationary satellites were beaming back tropical images every thirty minutes. Today satellites and reconnaissance aircraft measure (or estimate, in some cases) air pressure, temperature, humidity, and wind speed throughout a storm and the surrounding atmosphere. Forecasters do have a clue about a storm's destination—a very good clue.

Benjamin Franklin, of all people, may have been the first of us weather watchers to figure out that the big storms are independent phenomena with somewhere to go. I read the story in *Hurricane Watch*, by Bob Sheets, a former director of the National Hurricane Center, and Jack Williams, founding editor of the famous *USA Today* weather page, which single-handedly changed the way weather is reported in the United States. On October 22, 1743, stormy skies over Philadelphia blocked Franklin's disappointed view of a lunar eclipse. Weeks later, he was quite surprised to read in the newspapers that Bostonians had enjoyed a fine view of the phenomenon *before* that city had been socked in by the storm. How strange, Franklin thought, because the wind in Philadelphia on the fateful night had been from the northeast—the direction of Boston. How could the storm *from* the direction of Boston have hit that city *later* than it had hit Philadelphia? As Sheets and Williams tell us, it was seven years before this wise man revealed, in a letter, his theory of independent movement. He called it right. The storms do move independently, but they are embedded in the atmosphere and track in conjunction with the overall movement of that atmosphere, which in turn is often guided by powerful steering currents.

A key controlling factor for storms is the jet stream, a current of fast-moving air in the upper level of the atmosphere, usually between six and nine miles above the surface. It is typically thousands of miles long and a few hundred miles wide, but only a few miles thick, and as it shifts now to the south, now back to the north, it denotes the strongest contrast in temperatures at the surface—a cold front, classically. In the northern hemisphere, another major steering force is the overall east-to-west flow *at the surface* in the tropical latitudes, over both the Atlantic and the Pacific oceans. We know

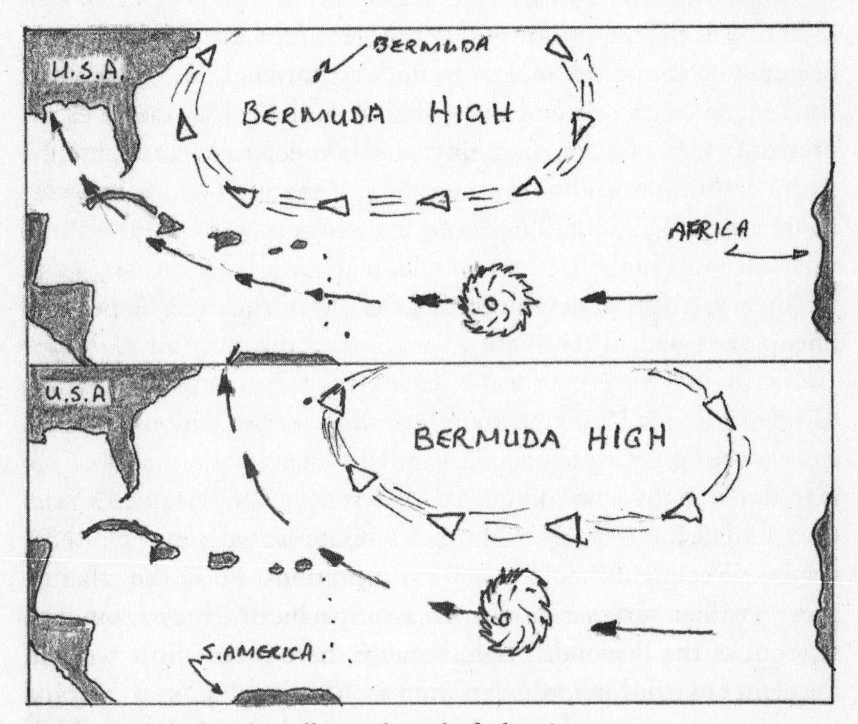

The Bermuda high and its effect on the path of a hurricane.

this flow as the trade winds, which sweep from the east and northeast across all of the islands of the Bahamas and the Caribbean.

Therefore it is essentially impossible for an Atlantic storm in the lower latitudes to move from west to east for any considerable distance or period of time.

The most important single steering factor in the North Atlantic is often the notorious Bermuda High, a zone or ridge of high pressure often parked—that's always the verb—over Bermuda, about seven hundred miles east of North Carolina. This midlevel subtropical ridge can shift east or west, north or south, and it can be larger or smaller than normal, stronger or weaker, but a high pressure of some sort is almost always parked somewhere off the Eastern Seaboard. By definition, such a high-pressure zone repels any low-pressure zone. Tropical storms forming off the coast of Africa may move almost directly north, passing harmlessly to the east of the Bermuda High, or they may head toward the west below it. What they will not and cannot do is plow right into that high-pressure zone. Often enough they track to the west, strike the U.S. coastline, then swing north and east in the higher latitudes and out over the North Atlantic. Or they may execute this sweeping loop around the high before reaching the seaboard. Or, if the high is situated relatively close to the U.S. mainland, the storm might be shoved into the Gulf of Mexico.

As it happened in 2005, three British friends of mine had started out in late June to cross the North Atlantic from Florida to the Azores in their thirty-five-foot sloop. I pitched in with remote weather routing. Using a marine Internet service that connects to ships via marine high-frequency radio signals, they e-mailed me each day with their latitude, longitude, and course. Each morning, I gathered all the relevant maps and satellite imagery and plotted a suggested course to maximize wind conditions. I also sent them a short weather summary and forecast. Specifically, they wanted to stay out of the Bermuda High, because there is very little wind in the center of that high, which is not much fun on a sailboat. Nor did they want to sail too far north and get caught in one of the low-pressure systems that spin off the Canadian landmass every week or so—definitely not much fun on a sailboat.

It's amazing how great communications at sea have become in recent years. On my cruise across the South Atlantic in '89, there was no Internet, and I couldn't afford any of the rather expensive and bulky marine shortwave radio systems, much less the satellite gear. So all we had was marine VHF radios, and even though the antenna was fifty feet high its effective range was about thirty miles, occasionally fifty miles. One day, however, that was good enough, as a massive supertanker heading for the LOOP oil facility just off the Louisiana coast overtook the little sailboat at 10 degrees west, 10 degrees south, as I recall. The skipper called us up on the VHF, wanting to know how we were doing and if we needed anything. He even offered us some steaks, but we'd just caught a nice mahimahi and said no thanks. He also offered to make some calls on his satellite telephone. What a surprise for our families to get this call from a rather British-sounding Captain Robertson of the SS *Leonia*, assuring them that the *Maggie* was within sight and all two aboard were doing well. This good captain took some photographs as his supertanker plowed past and kindly forwarded them months later. I still prize the picture of my well-trimmed little honey sailing along in the middle of the Atlantic Ocean.

As the Hurricane Katrina emergency unfolded, due in good measure to the lack of interoperability across the communications net—simply put, the first responders couldn't talk to each other—I thought about the ease with which I'd routed my friends across the North Atlantic, the ease with which Captain Robertson had made those satellite calls sixteen years earlier. But in 2005, in Louisiana and Mississippi, in the United States of America . . .

My friends' safe crossing to the Azores in June took about six weeks. For those same six weeks I watched the Bermuda High and everything else going on in the North Atlantic. The weather was great for the sailors, generally speaking, but what I saw was alarming as it related to hurricanes. The year before, four had hit Florida in a matter of weeks. The Bermuda High was pretty stable during that period—a big reason all of these storms tracked up and over the Florida Peninsula. Now, in 2005, the high seemed to be mostly west of its parking space the year before, suggesting that storms taking the southern route around this high would not—could not—start the turn to the north until they were deeper into the Gulf of Mexico. And so it had happened. On July 4, Hurricane Dennis formed up in the eastern Caribbean, tracked to the northwest, hit Cuba as a powerful Cat 4 storm, weakened but then reintensified shortly after entering the Gulf on July 6, then hit the Florida panhandle as a Cat 3. On the day that Dennis entered the Gulf, Hurricane Cindy hit southeastern Louisiana with 70 mph winds and a lot of rain. This storm had popped up in the northwestern Caribbean, crossed the Yucatán peninsula, and then moved straight north. (Initially categorized as a tropical storm at landfall, Cindy was upgraded to a minimal Cat 1 six months later. Ironically, Michael Brown, the soon-to-be-infamous head of FEMA, chose August 23, the day tropical depression 12 was born, to announce southeastern Louisiana's eligibility for federal disaster funds arising out of Cindy's floods.) And then came Arlene, Bret, Emily, Gert, and Jose—storms that had already crossed the Gulf of Mexico that summer, heading one way or another.

The NHC forecasters expected the new depression to move slowly northwestward toward a weakness in the Bermuda High. However, all of the computer models forecast that the high would fill the weakness and repair itself, in effect, within thirty-six to forty-eight hours. This action should drive the new depression westward across southern Florida and then into the eastern Gulf of Mexico within ninety-six hours—four days. It was a relatively easy call, and even with the first bulletin I thought, here we go again. The summer of '05 seemed to be our turn along the Gulf Coast. Hurricane activity tends to be cyclical anyway. The 1950s and 1960s were relatively active in the Atlantic Basin; the 1970s and 1980s a little less active; then business began to pick up again in 1995. Since that year, forty-two major hurricanes (Cat 3 or stronger) have developed over the Atlantic Ocean. To find a period remotely as active we have to go back almost fifty years, and there are reasons to believe this current active era could last awhile longer.

On Wednesday, depression number 12 could boast of sustained

surface winds of 39 mph, and it had therefore earned official tropical storm status. The previous one had been Jose, so this name would begin with K. That's all I knew (the lists are posted years in advance, but I don't study them) until 11:00 A.M., when advisory # 4 christened her Katrina. I know a French TV documentary producer (also a sailor) with that name, and she's pretty feisty. This was the image that came to mind when I saw the new name—a storm with lots of wild spirit. But I guess that's always the case with a storm. South Florida could expect landfall on Thursday, with an almost certain track across the state and then into the Gulf of Mexico.

At home, this nervous sailor started to check out the anchors, rodes, and shackles stored in the attic. I warned my wife, Lorie, that the track of this new one looked ominous. Let's start to get ready. Our well-rehearsed plan calls for loading the emergency gear on our small motorboat, then taking it and the *Maggie* from the marina just off Lake Pontchartrain up to my chosen hurricane hole on the Tchefuncte River, close to a cypress swamp. This plan also requires trips to the gas station, because we'd need lots of fuel for the cars, the motorboat, and the generator. (In Livingston Parish, with its wealth of forests—I say "wealth," although the urban sprawl is quickly decimating these wonderful stands of trees—we lose power with any kind of wind storm. The small generator has been a good investment.)

As an outdoors-type person who isn't outdoors often enough, I don't need much of an excuse to check out the weather and the weather maps, and during hurricane season, none at all. After putting a cup of English tea in the microwave, the first thing I do every dawn is to open any e-mail alerts from the National Hurricane Center. Then I go to the NHC Web page and review the five-day track forecast and the discussion paragraphs. I'm not a meteorologist, but I follow their technical talk easily. Next I go to HurricaneAlley.net, a private-subscription provider with excellent graphics, including the "spaghetti plot," which shows all of the predictions of the dozen different computer models, including some from overseas, superimposed on one map. This graphic can give me or anyone else an idea of how much confidence to place in the official five-day track forecast.

General agreement among the models gives me some confidence in the forecast track, but if the spaghetti plot lives up to its name, with the various forecasts showing a lot of spread, I may go to other Web pages, especially a site supplied by the Navy, and to the marine weather charts that are transmitted through the marine WeatherFax service from Coast Guard stations, also available on the Web.

In this way I formulate my own opinion about a storm, because I like to give our storm-surge modeling team at the Hurricane Public Health Center a few days lead time before we go operational. There are six of us on this team: a computer modeler, an oceanographer, a GIS expert, myself, and two graduate students. The computer whizzes needed two years to set up our computer model specifically for Louisiana and to adapt it to the requirements of LSU's SuperMike supercomputer (named after either LSU's mascot Bengal tiger or former governor Mike Foster, who got the funding for the blazingly fast machine). I wouldn't want to say these storm-surge models are the most important work we do, but they are certainly the most visible to the public. They have put us on the map, that's for sure. In the case of Katrina, on Wednesday I advised the team and some people with various agencies that we would probably go fully operational either late Sunday or early Monday. This would be the signal to the handlers who groom SuperMike to alert other users they could be bumped from the queue. Our models need a lot of computing horsepower.

Katrina was expected to strengthen modestly, and that's just what happened. She was a Cat 1 storm with sustained winds of 80 mph when she hit Florida between Miami and Ft. Lauderdale at 6:30 P.M. Thursday and swept quickly across the peninsula. That same afternoon I had been on a conference call with EPA officials and some folks from the state concerning the bill introduced in Congress by Louisiana senator David Vitter that would allow—no, *encourage*—the mining of cypress trees. These cypress resources have been logged one time—the early 1900s—with predictably bad results. But cypress mulch is in great demand these days—for gardens—so the squeeze is on from folks who know—every one of them—that the cypress swamps are some of the best defense against storm surge we have in Louisiana. At the state level we wanted to make sure the EPA understood this value and therefore why this bill was so patently ridiculous. I couldn't believe we were even discussing this plan for desecrating the cypress swamps just as another hurricane was apparently on its way into the Gulf of Mexico, and I said so during that conference call.

But what now for Katrina? Thursday night is when it got really interesting for the forecasters and the rest of us, because the meteorological certainties that had propelled the storm across Florida fell apart; the Bermuda High now had competition for influence. At all times across the entire globe, areas of high and low pressure are scattered throughout the atmosphere, and all of them within hundreds of miles will play some role in the track of a hurricane—the butterfly effect, the chaos effect. Nor are these pressures and steering winds uniform at every altitude. Experience and computer simulations have taught us that the strongest steering currents of all, in most cases, are those between thirteen thousand and sixteen thousand feet, but others have an impact as well. Nor are the large storms entirely passive regarding the influences from the surrounding atmosphere. They influence the air mass in which they're embedded. Sometimes parsing the net effect of the numerous influences is pretty straightforward, and most or all of the computer models are in general agreement. Sometimes, though, they disagree wildly and the spaghetti plot is just that. Just one particularly large and powerful cluster of thunderstorms on one edge of the storm can apparently cause significant jogs in the track. How can any computer program account for such uncertainty? It can't.

Steering currents can also be so weak overall, or in such well-balanced conflict, that the storm dawdles, loops this way and that, doubles back, meanders. This was the case in 1960, with the very powerful Hurricane Carla in the Gulf of Mexico before she finally came ashore with ferocity in South Texas. (Carla is well known in broadcasting circles as the storm that gave us Dan Rather, who made his name on the national scene while reporting live from the seawall in Galveston for the local CBS affiliate.) In June 2001, the track of tropical storm Allison provided devastating evidence of what can happen with even a modest storm when no single steering influence dominates. Some researchers argue that Allison wasn't even a tropical storm, technically speaking. Regardless, this rainmaker caused many billions of dollars in damage from flooding—by far the most costly tropical storm in this nation's history—as it made landfall on the Texas coastline south of Houston, floated northward for about two hundred miles, stalled, floated back to the Southwest and returned to the Gulf at almost exactly the same spot it had come ashore four nights before. Then it made a hard left, hugged the coast toward the east before coming ashore again in central Louisiana, proceeded east-northeast directly across Lake Pontchartrain, across the southern states and up the East Coast—with bad flooding all along the way. By and large, the forecasters saw the problem shaping up, because the storm was tugged this way and that by competing subtropical ridges, one east of Florida, one west of Texas.

In Allison, a measuring station northwest of Houston recorded twenty-six inches of rain in ten hours; the Bayou City's bayous filled to record levels, then overflowed. One small section of the Louisiana coast received over twenty inches of rain. These are monsoon-type numbers. My neighborhood in Livingston Parish took on twenty inches of rain, and all the local creeks flooded—predictably, I should add, because the urban sprawl swallows the cheaper parcels of land—the upland wetlands—and the "concrete effect" becomes more pronounced. This term, I believe, is almost self-explanatory. As the sponge and water-soaking values of wetlands are replaced by the 100 percent runoff, antivalue of concrete, the flooding impact of any amount of rain necessarily gets worse. In some parts of Louisiana, the hundred-year flood level has been redefined more than once and is now more than two feet higher than it was just twenty years ago. My home is on a mountain, relatively speaking—forty-one feet above sea level—and it was almost flooded by Allison's rains. Without four-wheel-drive we would have been utterly marooned for three days. I have tried my best to inform the local drainage board about these perils, but it is obvious to me that developers have a stronger voice in these matters. Now that the Katrina tragedy may get the government to rethink its coastal wetland options in Louisiana, why not also take a lesson from Allison and rethink the destruction of *upland* and river basin wetlands? But I doubt this will happen. The mighty greenback drives a lot of bad decisions. (Here's a tip to all prospective home buyers in Louisiana: If the road on which you drive to reach the subdivision is elevated on its own seemingly innocuous little earthen levee above the surrounding terrain, the highway department may know something you need to know. Ask the developer some hard questions about the flood plain in this vicinity. Regardless of what you're told, you might think about building your house on stilts.)

On Thursday night some of the NHC computer models had Katrina turning quickly to the north and hitting the Florida panhandle with a tolerable blow, but others were scarier. They had it proceeding west into the Gulf, strengthening, and only then swinging to the northwest and threatening the coastlines of Alabama . . . Mississippi . . . perhaps even Louisiana. The main source of their disagreement was a low-pressure center over the Midwest, which would try to pull this powerful low pressure in the Gulf of Mexico toward itself. But pull how hard and how fast? Those were the questions, and I had more than a hunch about the answers. Earlier in the year weak Cindy had moved over New Orleans, Dennis off to the east, Emily to the west, Franklin, Harvey, and Irene north into the Atlantic, and Gert and Jose into southern Mexico with little punch. But now the Bermuda High looked strong and the Gulf looked weak, and I decided Katrina was coming our way. Plus, the Gulf was so warm fishermen had been joking that there was no need to cook the catch, even if you didn't like sushi. They were pulling the fish out of the water prebaked. After a meeting Thursday at the Louisiana Department of Natural Resources to share some ideas about mining offshore sands to restore our barrier islands, I had told the assembled group, "This one could have our name on it."

By Friday morning, the computers were providing corroboration. The Bermuda High was pushing Katrina well into the Gulf.

The spaghetti plot was sorting itself out. This lady was headed west and then, presumably, northwest, and then, presumably, north. Some strengthening was expected. So it was on Friday morning that the e-mails and phone calls started flying in the offices of the LSU Hurricane Center and the Hurricane Public Health Center. "Looks like we might be busy the next few days," Marc Levitan, the director of the hurricane center, wrote the team by way of classic understatement. Marc asked if we were ready to start running the storm-surge models. Without a drastic change of course, we'd start the new modeling on Saturday morning, with the first publication on our Web site that afternoon. I chimed in with a reminder to the users of that site not to broadcast the Web address to the public, because the ensuing demand would surely overwhelm our underpowered server and down we would go, slowly but surely. This had happened during Hurricane Ivan the previous year, when downloading just one of the animated movies of the surge models had required one hour or more instead of the usual three minutes. There were complaints. So just paraphrase the results this time, I begged. (My appeal actually worked. Our computer was able to keep up with the demand.)

The National Hurricane Center's official advisory for 11:00 A.M. Friday stated that the hurricane, still Cat 1, could become a Cat 2 by the following day. Wrong! It was a Cat 2 within hours, and the 5:00 P.M. advisory stated that now Cat 3 strength was possible by Saturday. As difficult as forecasting the storm track may be, forecasting intensity is manifestly and inherently more complex and fraught with the possibility of error. A notorious example among forecasters was Hurricane Kenna, which formed off the Baja peninsula of Mexico in late October 2002. Kenna was supposed to make landfall as a moderate storm. In fact, it became a Cat 5 storm and then hit the central Mexican coast as a Cat 4. This forecasting gaffe was the subject of much discussion at the national hurricane conference held in New Orleans the following April. As researcher James Franklin said, with a bit of exaggeration, perhaps, "We were only off by 110 mph. It would be hard to miss by more than that."

Kerry Emanuel, a researcher at MIT and the author of *Divine Wind*, a terrific coffee-table book about hurricanes, has laboriously compared the actual wind speeds of several hundred hurricanes and typhoons over the past forty years with the theoretical wind speed for that storm as determined by the theory of heat engines. Surprisingly, only 10 percent of the storms generated winds approaching 90 percent of the theoretical maximum. A majority did not generate winds even half as strong as they were theoretically capable of. In short, very few storms get the most out of themselves.

It's hard to become a storm, and apparently it's hard to reach maximum potential as a storm. "Something is holding them back," Emanuel writes. "Meteorologists suspect two main culprits." The most important of these is believed to be wind shear, which I'll define in simplest terms as any conflict between the direction of air flowing into the center of the storm at different altitudes. A wind coming from the east at five thousand feet and one from the west at ten thousand feet creates a lot of shear, which will probably have a weakening impact on the storm. Such conflicting winds may disrupt the symmetry of the counterclockwise circulation. They may also import cooler or drier air. The wrong strong winds aloft—thirty thousand feet, say, five or six miles up—may shear off the tops of the highest thunderstorms, definitely disrupting the efficient functioning of the heat engine. Decapitation, we call it.

Shearing factors are deemed one key reason the South Atlantic is the only equatorial ocean that produces almost no tropical storms at all. Those prevailing westerlies are closer to the equator than they are over the North Atlantic, shearing apart impending storms with great efficiency. In addition, the cold Benguela ocean current sweeps up the west coast of Africa, so the waters feeding into the equatorial current just south of the equator are colder than their northern counterparts. The only exception to the rule within memory developed off the coast of Brazil in 2004. The Brazilians didn't even want to call this storm a hurricane, but our NHC was certain and therefore gave it a name: Catarina. Some climate-change scientists believe its appearance is a reflection of global warming. In a world made warmer by increased greenhouse gasses, the South Atlantic could be in for more tropical storms.

The second factor that apparently saps hurricanes of strength is the fluctuating temperature of the water over which they pass. The warmer the better for strengthening, of course, but the surface temperature of large bodies of water—oceans and their gulfs—will vary by several degrees. This doesn't sound like much, but small variations can and will make a big difference. I said that Katrina, on her journey across the Gulf of Mexico, generated energy equivalent to one hundred thousand atomic bombs. That's amazing, but how about this one: A temperature increase of just 1 degree Fahrenheit in the Gulf could generate the energy equivalent of *one million* atomic bombs. So temperature matters. So does depth. The topmost, warmest layer of water can be quite shallow—in much of the Gulf of Mexico, less than fifty feet. A powerful storm that churns the water enough to bring cooler water to the surface is thereby choking off the heat that it requires to grow or sustain itself. A hurricane will sometimes leave a trail of cooler water in its wake, which can be discerned by satellite imagery. It is not at all coincidental that hurricanes often strengthen over the western Caribbean, where the warmest layer of water can be over three hundred feet deep. It was here that Hurricane Wilma metastasized from a Cat 1 to a Cat 5 hurricane overnight (October 19, 2005), recording the lowest barometric pressure ever in an Atlantic Basin storm. Wilma then predictably weakened but still pounded the resorts of Cozumel and Cancún on the Yucatán peninsula before turning sharply northeastward and hitting the southwestern coast of Florida—all according to the forecasters' plan.

Had Wilma proceeded straight north into the Gulf of Mexico, she might have found the warm waters of the Loop Current conducive to maintaining her record strength. This odd feature of the Gulf has presumably played a critical role in the development of many storms. The Loop is, in effect, an extension of the famous Gulf Stream, whose main current flows north through the Yucatán Straits from the Caribbean, then swings immediately to the east, past the Florida Keys and up the Eastern Seaboard. It is a closed eddy that sometimes extends farther north into the Gulf of Mexico. Like the Gulf Stream itself, it is about twelve hundred feet deep, deep enough to prevent cooler water below from rising to the surface as the winds churn the seas. Some researchers have surmised that Hurricane Camille, which struck Mississippi in 1969 with winds of almost 200 mph, the highest hurricane winds ever recorded on the continent, must have proceeded right up the spine of a perfectly positioned Loop, ingesting new energy every mile of the way. On the other hand, that can't be the whole story with Camille, because other storms have passed over the Loop and not strengthened at all. In 2005, Katrina certainly benefited from the Loop, which extended almost all the way to Louisiana in late August.

For the computer modelers, however, the basic problem with water temperature is that the specific number is simply not known for wide areas of the ocean at different depths, and it is financially and logistically infeasible to sow the thousands of bathythermographs necessary to collect this data. The utility of satellites is limited, because only the first few meters of the ocean reflect light or infrared waves. Coming on-line right now are exquisitely sensitive altimeters that can measure the level of the sea to within a few inches, thus providing information about the thermal structure beneath, but, by and large, the computer modelers have no choice but to use broad assumptions about temperature based on prior experience, time of year, and whatever measurements are at hand. Theoretically, a slower storm will churn up a given area of water more completely, therefore more likely bringing cooler water to the surface and hindering any strengthening, but can you reliably account for this factor in a computer model? It's tough. Researchers often resort to SWAG, or "scientific wild-ass guess." And even if completely accurate temperature data were available, no one can pin down with precision the exact impact of water temperatures on the overall system. Therefore the predictions are often fraught with error.

Like Kenna off the Baja peninsula in 2002, the puzzling case of Hurricane Lili the same year demonstrates just how tough predicting intensity can be. Twelve hours from landfall, NHC forecasters predicted a strike on the central Louisiana coast as a Cat 4 storm pushing a twenty-foot surge. Our LSU storm-surge models, which had just come on-line, predicted thirty feet of surge and, obviously, serious flooding west of the Atchafalaya Basin. But then Lili came ashore as a much weaker Cat 1, with a surge of only eleven feet. In the post-mortem, the forecasters studied a jet of drier air moving into the region—and therefore into the storm—from the west. They had believed this drier air would be entrained into the storm, but they underestimated its impact. A second factor was probably the passage of Hurricane Isidore across the same section of the Gulf just a week earlier. Isidore had seriously weakened over the Yucatán and was only a tropical storm at landfall near Grande Isle, Louisiana, but he left stirred up and presumably cooler waters in his wake. In addition, runoff from Isidore's inland rains would have still been pouring into the Gulf along the Louisiana coastline—water several degrees cooler than the main body of water. Only in retrospect did all become clear—or clearer.

As for Katrina, her strength exceeded expectations almost by the hour on Friday, and with each passing hour the excitement mounted for hurricane watchers all along the coast. This is an interesting aspect of human nature, of course: the bigger the storm the worse the damage, but the bigger the storm the greater the thrill. For professionals, the biggest storms get the juices flowing. That's a fact. You can see it with every weather forecaster on television. They want a monster, and they're disappointed every time the beast drops a notch in the category rating. It's human nature. In our various offices at LSU, calls poured in from around the state. What did we think? They always seem to veer away—what about this time? Is it really going to intensify? Could this be *it*?

Yes, this could be *it*. We could be in big trouble on Sunday night, maybe Monday morning. That was my (and everyone else's) conclusion by early Friday afternoon as I studied the spaghetti run on the HurricaneAlley Web site (not much spaghetti, lots of agreement), checked the weather sites, and considered the National Hurricane Center's five-day forecast. I told the SuperMike team that starting the following morning we would need access to the full array of processors, and I told the Federal investigators on a landfill matter with which I was involved, who were due in court for a hearing on

Monday, that they would be picking up bodies instead. Governors Kathleen Blanco of Louisiana and Haley Barbour of Mississippi declared states of emergency. The state Homeland Security officials in charge of the emergency operations center in Baton Rouge prepared to go operational. Late that night at home I loaded the motorboat with the anchors and the rodes. The next morning Lorie and I would roll out for the hurricane hole on the Tchefuncte River.

TWO

LEAVE, PLEASE!

At 4:00 A.M. Saturday morning, the track for Katrina in the Gulf of Mexico still looked bad. Nothing good for New Orleans had happened over the previous five hours, and nothing good was expected to happen over the next five or ten. If this storm didn't hit the city dead on, it was probably going to come awfully close. I called Hassan Mashriqui, our storm-surge modeler. Was he ready? He was ready. Of course. Mashriqui (we always call him by his last name) knows full well the importance of these surge models.

In its simplest terms, the storm surge is the vast mound of water generated by the high winds of the storm (there are other factors, but the winds are the main one) and pushed across the open sea, and with most storms this surge accounts for the most catastrophic damage. (Andrew in Florida was the outstanding exception in recent years: that was all wind.) In tandem with the models used for forecasting hurricanes, the ones that predict the surge have revolutionized our understanding of storms and our ability to prepare for them.

From our perspective at the LSU Hurricane Center, the story of the surge models began in the early 1990s, when I worked with two excellent scientists and great guys, Paul Kemp and Joe Suhayda, on some early computer modeling designed to predict the synergistic benefits of rebuilding our barrier islands in tandem with our wetlands, all in order to save our state. Paul and I had been graduate students together in the since disbanded Department of Marine Sciences at LSU. That was in the late seventies and early eighties. Both of our projects—mine on the evolving Atchafalaya Delta southwest of New Orleans, his on the chenier coast to the west—were very field-time demanding. We spent thousands of hours out on the marshes, in the brutal, humid heat of the summer and on winter days that can be just as brutal on the other end of the scale. (This fact really surprised me, my first year on the scene. I would never have imagined that the Atchafalaya in winter is just above what we call "melt ice" temperature. It's cold water.) Paul is incredibly smart, both as a scientist and as someone who can read a situation. He also has a heart of gold and is willing to fight for the underdog—my kind of guy—and since we are both about the same size and stature and have fair, curly hair, we've been referred to as "the terrible twins." So be it. I'm proud to stand side by side with this comrade-in-arms.

Coastal engineer Joe Suhayda directed the Louisiana Water Resources Research Institute at LSU. A longtime faculty member, Joe was one of the very first of us to sound the alarm about the peril of New Orleans, and he had the great idea of taking film crews into the French Quarter, where he would hold up a long surveying rod to indicate how high the water would be in the worst flood—way past the wrought-iron railings of second-story balconies. (Joe has also put forth the idea of creating a "community haven" with a two-story wall that would seal off the French Quarter, downtown, government buildings, a hospital, and some housing. I have never been a fan of this approach, and I believe its drawbacks have been precisely illustrated in the events of August 2005. The sections of the city that would be protected by Joe's wall are, for the most part, the sections that stayed dry during Katrina. The exception is downtown, which had several feet of water during Katrina and would have been inside the haven. Yet look where New Orleans is at this writing: culturally and economically crippled. I think we have to protect the whole city.)

In the early nineties Joe, Paul, and I were having a problem convincing "the agencies" of the value of restoring the barrier islands. Some of those agency folks even argued that the barrier islands were not eligible for federal restoration dollars because they were not vegetated wetlands, an argument that incensed the three of us. They even invented some scheme to determine the wetlands benefit of proposed projects that totally ignored the reductions in wave impact provided by healthy barrier islands. They also ignored the reduction in storm surges and tides provided by healthy barrier islands. This was simply crazy. Not long before—August 1992—Hurricane Andrew, primarily remembered for the incredible wind damage inflicted on the towns south of Miami, with 125,000 homes seriously damaged or destroyed, had also ripped into Louisiana as a less dangerous but still potent Cat 3 storm. Fortunately it struck a relatively sparsely populated area on the central coast, where the wetlands are their healthiest, expanding even ("prograding," as we say). Damages in Louisiana amounted to $1 billion ($35 billion in Florida, second only to Katrina). This was serious damage, but it could have been much worse. Thanks to the healthy wetlands the storm surge was only eight feet. Eight hours after Andrew's landfall, I flew the coast with well-known Louisiana coastal oceanographer Shea Penland. The damage to wetland areas under stress from bad management was much worse than the damage to healthy marshes. Anyone could have seen the difference.

So-called wetlands scientists should have been able to understand that the barrier islands also help the wetlands, but they didn't. Joe, Paul, and I wanted a good computer model that could prove the benefits of the islands. Surge models had been around for a quarter century, when Chester Jelesnianski of the U.S. Weather Bureau (as it was still called at that time) developed SPLASH (Special Program to List Amplitudes of Surge from Hurricanes). This model scored an immediate triumph, predicting the devastating surge that accompanied Hurricane Camille in 1969. Jelesnianski then developed SLOSH (Sea, Lake, and Overland Surges from Hurricanes), which is still in use by the National Oceanographic and Atmospheric Administration (NOAA) and other agencies. Joe Suhayda had been working with a FEMA model that the agency used to calculate flood risk for insurance rates, and we rounded up thirty thousand dollars from the late Terrebonne Parish engineer Bob Jones, who believed as we did: the islands matter. The main town down there is Houma, and the marshes to the south are as ripped up as any in the state. Our idea was to use the computer to compare what the tides did to these marshes with and without the barrier islands offshore. If the models showed what we thought they would, we could really create on-the-ground value for restoration. And the models did show this value. Irrefutably. With happy faces we took the work to a meeting in Lafayette, where I, at least, was amazed to confront a group of agencies who were proudly armed with a long treatise about why computer modeling was no good, period. The U.S. Fish and Wildlife Service led the way in this obstructionism. I really thought I was in the Third World—no disrespect intended toward those nations. South Africa, for one, was light years ahead of us in terms of computer models. This attitude in Louisiana, in the United States, in light of SPLASH and SLOSH, was ridiculous. I got quite angry and frustrated, but Paul Kemp showed his mellow mettle yet again, stepped into the foray with his typically pleasant, disarming demeanor, and after an hour had convinced the agencies to take a fair look at our work.

That was the first use in Louisiana of surge models. Two years later, in 1994, then-governor Edwin Edwards appointed me to run the state's coastal restoration program, a story I'll relate later as part of the much larger story of the checkered history of wetlands protection in our state—an all-important subject if you want to understand how Katrina clobbered New Orleans. As was the case so often in those days, the team from the Environmental Protection Agency understood which projects made ecological sense, and restoring barrier islands definitely made the grade. Most of the academics also understood, but the U.S. Fish and Wildlife Service was still against us, and so was the Army Corps of Engineers, because we were the new guys with the new agenda. I wasn't going to let their obstruction stop me. I was going to make sure science got into the picture, and science, have you heard, *uses computers.*

The coastal restoration program could "sole source" contracts up to about fifty thousand dollars, so Paul Kemp got money to further develop a river basin model; Joe Suhayda got funding to get his first surge models operational; and a young Irish scientist at LSU, Gregory Stone, got some funding to set up a wave model for the Louisiana coastline. A decade later, modeling has become the rage, as well it should have. I like to believe that being a little hardheaded and at the same time resourceful paid dividends. I'll even argue that the funds I sole sourced to those three scientists may have been the best investment the state government has ever made. (Of course, some jaded observer of the local political scene will quip that the competition hasn't been very tough.)

In 2001, when I was trying to secure the funding for LSU's Hurricane Public Health Center, I cast around for what I thought would be the best surge model for both planning and operational support. I knew we needed this element in our package of features. Joe Suhayda was getting ready to leave LSU, and his low-budget model was neither the latest nor the best. I soon decided that the latest and the best was the ADCIRC model developed by Dutch-born Joannes Westerink, a professor of civil engineering at Notre Dame, and Rick Luettich, of the University of North Carolina. Both are geniuses at computational fluid dynamics, and their model is, for our purposes, without a peer. The acronym stands for "advanced circulation." The other widely used model, SLOSH, is excellent for its purposes, but somewhat limited when compared to ADCIRC, which works at a higher resolution, in effect—an almost infinitely fine resolution in theory, and much finer in fact. This is an important advantage when dealing with areas such as New Orleans, where small-scale features can have huge consequences, as we shall see. ADCIRC is also much more adept at simulating convoluted shorelines and incorporating features like highways and canals that can block—or accelerate—storm surge. ADCIRC can include tides (though it doesn't for our predictions right now); SLOSH cannot.

ADCIRC is a vastly complex mathematical and computation engine—just as complex as the programs for predicting a tropical storm's course and intensity. I am not a modeler, but I can appreciate the beautiful science. The equations in the model work their magic with data pertaining to the storm itself—position, track, wind speed at the surface (extrapolated, in most cases), and barometric pressure—and to the bathymetry of the seafloor and the topography of the floodplain (including bays, rivers, and significant obstacles). From a numerical standpoint, things get very sticky as the water front advances and retreats. The model must "wet" areas that start out dry, and vice versa. The water coming down rivers such as the

Mississippi and the Atchafalaya is a considerable factor. Vegetation is very important: Obviously, densely packed cypress trees offer more resistance to the surge movement than low-lying salt marsh grasses, and these grasses offer more protection than open water, which offers none.

Most of the model's data is organized in a grid that establishes the computation points. (An equivalent computation grid is set up for the models that forecast storm track and intensity.) In the surge model, the grid over the open ocean might be 25 kilometers (15.5 miles) on edge; that is, a computation point every 25 kilometers. On the continental shelf, the resolution gets finer as the grid tightens to 5 to 10 kilometers. In the most populated areas of the Louisiana floodplain, it tightens even more, getting as small as 100 meters, a level of detail that is critical for us, because we depend on both natural and artificial levees (including raised roads and railroads) for defense against the surges, and we can account for these with ADCIRC. By comparison, Joe Suhayda's old model worked with a grid only 1 kilometer (0.6 mile) on edge.

Our current model, ADCIRC S08, yields a grid with exactly 314,442 computation nodes. S10, under development, will have a finest resolution of 60 meters, for a total of 602,254 nodes. S20 will have a resolution of just 20 meters, or 2.7 million nodes—quite a number-crunching challenge. The modelers use hindcasting to perfect the program. Dozens of flood gauges are positioned throughout southern Louisiana, accurately monitoring the exact water height at each one. Location by location, ADCIRC compares the actual record with the computed, predicted level. Discrepancies are investigated and the explanations incorporated into the program. The programmers learn from the mistakes and improve the program.

How many lines of computer code run ADCIRC? I'm told that this common question can't really be answered with a single number, but when pressed to produce one that won't be totally misleading, just for the benefit of a colleague writing a book for the general public, our modeler, Hassan Mashriqui, answers—guesses—fifty thousand. This code takes an official advisory from the National Hurricane Center, unites it with the stored bathymetry and topography data, and produces a brilliantly colored graphic representing the expected surge for the particular advisory. That's the gist of it. The simulation takes two and a half hours on SuperMike, one of the world's fastest supercomputers, but the modeling team requires at least five hours to display the results on the Internet, because of all the pre- and postprocessing of data. (The potential for providing operational support during hurricane emergencies was one of the main reasons cited to legislators by Governor Mike Foster for building SuperMike.)

We send the ADCIRC outputs to a large listserve of emergency management officials at every level of government, nongovernmental organizations (NGOs), and the media. A version of the model is also used by the Corps for designing its levee system. (More to come on this subject, of course.) The Louisiana Department of Natural Resources uses it for studying coastal restoration projects.

As with all computer programs, the old adage "garbage in, garbage out" pertains to ADCIRC. Once we had the model, we needed the team to put it into effective action for Louisiana. I call these folks the "LSU surge warriors." First on-board was the main builder, Joannes Westerink (working long-distance from Notre Dame), then Paul Kemp, our oceanographer, and Mashriqui, whom I've known since he worked as a graduate student under Paul. I was immediately impressed by the polite fellow from Bangladesh with a very thorough understanding of computers, both software and operating systems, as well as hydrologic engineering, his academic background. Like Paul, Mashriqui has a heart of gold and is willing to fight for the underdog. Over the past decade, he, Paul, and I have worked on numerous projects together, including the saga of the Wax Lake weir, yet another aspect of the important wetlands question I'm holding for future discussion. The minute Mashriqui got his Ph.D. I hired him as an adjunct assistant research professor in the research center.

Then we have graduate student Young Souk Yang from South Korea; Ahmet Binselam from Turkey, an absolute whiz at geographical information systems; George Eldredge, a former student worker of mine, now a computer tech in the College of Engineering; and Kate Streva, a research associate who makes everything happen, and who was pregnant in her third trimester during Katrina. (Edan is the beautiful baby's name.) Early on, we also got great help from Jesse Feyen, one of Joannes's former students, now with NOAA. Tom Berg, whom I've never met in person, is also a fine friend. Tom runs the HurricaneAlley.net Web page. In return for free access to that subscription service we provide him with our surge-model outputs. In this way we get around the problem posed by our small server and can reach beyond the emergency management professionals who are our main "clients." In past years, when we needed more detail on model outputs and tracks, Tom always went out of his way to send those.

Notice that about half of the surge warriors are not American-born, and no surprise. This is where we're at in this country today. More and more of the faculty and the majority of graduate students in engineering and physical sciences are from foreign soil, and not just at LSU, everywhere in the United States. American kids are not that interested in long, demanding graduate programs that don't guarantee riches in the end. I think MBA programs may be sucking more than their fair share of the best students away from the sciences and engineering.

All of the surge warriors work on "soft money" (short- and long-term grants and the like), which makes us second-class citizens in the academic world, with no prospects for tenure. I'm jumping ahead here, but on day four of the Katrina emergency—Thursday—an important administrator at LSU told us that the operational support we were providing an array of agencies was okay, but what really turned him on was *federal dollars*, especially from the National Science Foundation. These make him really happy. I was flabbergasted at his timing. None of us in the room had had much sleep for a week (and would not get much sleep for many more weeks). No one was getting paid for providing this operational support. No one had been paid during the many previous emergencies. One of our many tasks was mapping 911 calls for the first responders, and on the next day, Friday, seventy-eight people were rescued from their attics based on our mapping. These were people who most likely would have died otherwise. And this guy was signaling that the university doesn't value what we were doing? There were some despondent and angry folks following that little pep talk. Marc Levitan and I had to go around and lift some spirits.

The man's general drift is also a sore point with me, as I've hinted earlier. Not that we don't need to study the umpteenth black hole a trillion trillion miles away. We do, but we also need the vision to see what society needs by way of practical science. A balanced view on research is lacking at many universities obsessed with competing for the big brownie points, where upper administrators' egos and boasting rights are more important than solving problems to the benefit of society at large. You would think that our work before, during, and after Katrina might have turned some heads at LSU, but not really. Witness the confession by the administrator. Witness the gag order placed on me by the school during the levee investigation a couple of months later (soon rescinded with apologies).

Before dawn on Saturday, August 27, I called Mashriqui. Were we certain that the university had cleared the way on SuperMike for the hurricane queue? Yes. The polite e-mail to that effect, addressed to all users, was time-stamped "2:32 A.M." (Someone else couldn't sleep, I guess.) Kevin Robbins, who runs the Southern Regional Climate Center and heads LSU's team at the emergency operations center in Baton Rouge, had e-mailed me minutes earlier asking that we run advisory #16, then #17. Kevin was as worried as I was about this storm; he wanted as much information as he could get. Mashriqui was worried. Everyone was—and everyone already seemed to be awake. The e-mails and phone calls were flying.

During the Hurricane Ivan emergency the year before, we had run two models at the same time, one using the track right down the middle of the National Hurricane Center's forecast cone of uncertainty, the other the track on the western edge of the cone. This was because the cone for Ivan kept sliding to the west as he churned across the Gulf, and we wanted to make sure that if the storm did take the most westerly track, we'd be ready with flood forecasts for New Orleans and vicinity. As it turned out, the models never showed flooding in the city, and there was never the fear in our offices that Katrina brought out. For Ivan, the city was evacuated based on the advisories coming from the NHC. In the end, after a long career of two-plus weeks in which it achieved Cat 5 status three different times—a record for an Atlantic storm—Ivan ended up hitting the Alabama shoreline as a Cat 3.

I felt that we would want to do the same thing for Katrina, running both the main track forecast and the western edge of the cone. We might also have more than one advisory running at the same time, so at 6:10 A.M. I sent another e-mail to the SuperMike folks: "This is an absolute emergency, we need all the nodes we can get, PLEASE." We got them. Katrina was starting to look like a monster; she had us in her sights. Even the computer geeks sensed this! Before long I e-mailed the NHC, because I had heard that they have a good draft of the advisory ready about an hour before it's released to the public. If they could e-mail that early draft to us, maybe we could save an hour. They didn't respond immediately—they were superbusy themselves, of course—and later I learned that they wouldn't have wanted to give us that draft anyway, because they make changes right up to the last minute.

Advisory #17, issued at 10:00 A.M. Saturday morning, put the center of Katrina 200 miles west of Key West, 405 miles southeast of the mouth of the Mississippi River. Movement was westerly, with an expected curve to the west-northwest. Sustained winds were approaching 115 mph—Cat 3—with strengthening anticipated. Cat 4 status was likely—that is, winds over 130 mph. At the office the e-mails and phone calls and media requests *poured* in. Who can catch AP? Who has time for NBC?! CNN wanted "people who can speak specifically to the issue of evacuation." Brian Kennedy, a producer for ABC, wrote, "I hope this note finds you well. We spoke last year around the time of Hurricane Ivan. . . . What may I ask will you and your team be doing in the next day? Are you available for an interview wherever you may be? We are in New Orleans." Okay, along with forty more that day, literally. The media had called us during Ivan, but now they really had our number. The drama was building, and maybe they could also sense the tragedy that was about to unfold. They asked, we answered.

When were the authorities in New Orleans going to evacuate the city?! The ADCIRC surge models were already scary. At 12:45 P.M. Mashriqui e-mailed the results of the first run, the one based on advisory #16, and added: "Note: Water gets in New Orleans from the Airport side. Very close and likely to overtop from every side." Ahmet Binselam immediately started to put the data on our Web page. Three minutes later I sent a note to Mark Schleifstein of the *New Orleans Times-Picayune*. There will be some flooding, I said. He was not surprised. In June 2002, the newspaper had published his and reporting partner John McQuaid's "Washing Away," a five-part series on the danger faced by New Orleans. The series won the Pulitzer Prize, and justly so. No one in any position of authority in the city of New Orleans or the state of Louisiana or the United States of America had any further excuse for underestimating what the city would someday face—and was now facing, in fact, barely three years later. (A decade earlier, the *Baton Rouge Morning Advocate*, as it was called before it dropped the "*Morning*," had raised the same alarm with its series "Ill Winds," written by Bob Anderson and Mike Dunne, which covered hurricane preparedness along the entire Gulf Coast.) I explained to Mark that he should ignore the flooding out west by the airport—in the model, a section of the levee in that area was lower than it actually is—but concentrate on the fact that the surge would be up to the tops of the levees, and there would be erosion and overtopping. Things were not looking good. Mark has a very matter-of-fact voice, but I could hear a slight rise in tone as he digested the news.

Shortly after three o'clock Saturday afternoon Mashriqui produced model #17, with the same upshot as the previous one. I immediately forwarded the results to every official at every level of government I could think of, and I added, "PLEASE NOTE. While the output does not show the city flooded our model assumes a uniform height for the levees around Lake Pontchartrain. . . . these surge heights will top the levees in some locations. In addition, we do not at present factor in the waves. . . . So please do not be swayed by the fact that the model does not show water in the city, it will get in. . . . PS the key is evacuate, evacuate." Landfall was about thirty-six hours away, maybe a little longer. I wondered why they hadn't ordered the mandatory evacuation of New Orleans. We were going to run out of time. Two hours later, in a note to Steve Lyons of the Weather Channel, I amended my assessment of model #17, pointing out that some levees assumed to be 14 feet above sea level are in fact only 11.5 feet. These will be overtopped, I told Steve. Pounding waves will induce erosion along the south shore of the lake—"a major factor." I concluded, "Basically, there will be a whole lot of flooding." When I'd met Steve earlier in the year I had asked him to always stress evacuation when reporting on any storm headed in the direction of New Orleans. In a new note to Mark Schleifstein of the *Times-Picayune* I wrote, "The bottom line is this is a worst-case scenario and everybody needs to recognize it. You can always rebuild your house, but you can never regain a life. And there's no point risking your life and the lives of your children."

Many days later we learned that Michael "Brownie" Brown, FEMA's head, wrote in an e-mail on Saturday, "This one has me worried." No kidding. Then Brownie expressed the wish that Jeb Bush were governor of Louisiana. What's really sad about this lame quip is that he may have been right, in a way. Just think. If someone with the clout of the president's brother had been the governor of Louisiana at some point over the past three decades, the levees protecting New Orleans and vicinity might be sound, and the swamps and marshes and barrier islands might have been restored with the thick grasses and plants and trees that serve as excellent storm buffering. That is to say, Louisiana might have seen something approaching the money that the federal government is pouring into the Everglades right this minute. The health of that famous river of grass is definitely important, but it's not life and death.

At 5:00 P.M. Saturday, Mayor Ray Nagin joined Governors Blanco and Barbour—and President Bush—by declaring a state of emergency. At the same time he issued a *voluntary* evacuation order, and added that it might become mandatory after his legal team determined the legality and wisdom of this act. One concern was the city's liability if they forced hotels and other businesses to close. I thought, Oh, come on, just do it! Nagin and everyone else in the business knew about the census number: 127,000 residents of New Orleans don't have cars. "This is not a test," the mayor said. "This is the real deal." He mused about the return of the swamp creatures, about a city underwater for two weeks, about a city that might never be the same again.

Mississippi county emergency managers, based on advice from the state, had already issued mandatory evacuation orders for their coastline, and parish emergency managers in Louisiana had done the same for the most vulnerable areas in Plaquemines, Jefferson, St. Bernard, St. Charles, and St. Tammany parishes. Residents had been leaving the imperiled lowlands farther south along the coast in moderate but steadily increasing numbers since Wednesday. They knew better than anyone that many of their communities were virtually certain to be underwater on Monday, no matter how Katrina might shift her track in the meantime. They didn't need storm-surge models to tell them this. The Mississippi River would become fifteen miles shorter. (It's true: Those last miles of the river's already ragged delta would be completely overrun by the surging Gulf of Mexico.)

Thirteen years earlier, on August 25, 1992, a mandatory evacuation had been ordered for the Morgan City region along the central Louisiana coastline only twelve hours before the arrival of Hurricane Andrew. The traffic jams were predictable; many people did not get out. Fortunately, they lived anyway. "It's a wake-up call!" emergency-preparedness officers proclaimed in the aftermath, and the following year the state legislature passed the Emergency Assistance and Disaster Act, setting up chains of command and coordination mechanisms for all levels of government. What happened? Six years later, in 1998, the evacuation of New Orleans for Hurricane Georges was a debacle. Six *more* years later, during the Ivan emergency, many of the half million residents fleeing the New Orleans area on Tuesday, September 14, needed ten hours for the eighty-mile run up to Baton Rouge.

I want to be fair to my state. Others have also had "colossal" traffic jams during hurricane evacuations—to borrow the adjective used in one of our LSU publications on the subject, referring to both Georges in '98 and Floyd the following year on the East Coast. During the Floyd emergency, evacuations were ordered for assorted coastal zones from Florida all the way north to Delaware, with as many as four million residents heeding those warnings and turning several interstates into parking lots. Colossal would also be the right word for the traffic jam in southeast Texas before the arrival of Rita, just three-plus weeks after Katrina. They brag about doing everything big in Texas, and they did this traffic jam big, too. The parked cars stretched from Houston to Dallas, 240 miles to the north. This may have been the worst tie-up in history. The photographs went around the world.

After Ivan, LSU transport engineers Brian Wolshon and Chester Wilmot, partially funded by our Public Health Center, used traffic computer models to develop a host of good suggestions to improve the contraflow mechanism. They worked closely with the state police and others to get it right. Maybe the most important advance was the realization that people in southern Louisiana need to be directed toward the *north*. During Ivan, the evacuations orders had started in Florida and rolled west with the storm. Floridians fled west and north, but most of the folks in Alabama and then Mississippi chose to head west, apparently. Near our house in Livingston Parish, Interstate 12 (an adjunct of I-10, in effect) was a parking lot, while the secondary roads had no traffic, as I learned when heading out and back from securing the *Maggie* in her hurricane hole on the Tchefuncte River.

Go north. That was the new idea, and it worked with Katrina. Contraflow provided nine lanes out of the greater New Orleans area. Folks leaving on Sunday did require sixteen hours to reach Houston three hundred miles to the west, but they had waited until the last moment, and the traffic was moving, if slowly. Computer models demonstrate that a good contraflow scheme increases outbound traffic by about 70 percent, but the models also confirm what all of us who drive cars already know: Any little bottleneck will screw things up, no matter how many lanes you have. This had been demonstrated clearly during the Ivan evacuation, when a work zone on Interstate 55, which intersects Interstate 10 about ten miles west of New Orleans, effectively negated the contraflow benefit on that escape route and contributed greatly to the problem. But what lengthy stretch of interstate anywhere does not have a work zone of some sort, on one side or the other? It's not practical to stop all work on the interstates in question during hurricane season. Automobile evacuation will never be easy, not when the forty-five million residents who now live in coastal regions from Texas to Maine are served by a highway system that is about as efficient as it was decades ago. The number of residents is projected to be almost seventy-five million by 2010, but the highway system will not expand correspondingly. Not even close.

There's another very major problem with evacuations: Most will prove to be unnecessary. Simply put, an effective evacuation takes time—at least forty-eight hours for a large metropolitan area confronting a major storm; in Louisiana, the preferred minimum evacuation time is seventy-two hours for a Cat 4 or Cat 5 storm—but the earlier the evacuation is ordered, the greater the likelihood for error in the forecast. Georges turned east, therefore New Orleans was spared. Ivan turned east, therefore New Orleans was spared again. Floyd eventually struck North Carolina, so South Carolina, Georgia, and Florida were spared. Rita turned east, therefore Houston and most of southeast Texas were spared. In each case hundreds of thousands, if not millions, of evacuated residents returned to undamaged homes and communities. In Texas, many thousands of evacuating cars were stuck in traffic just as forecasters were picking up Rita's turn toward the border with Louisiana. In fact, most of those who had evacuated on Thursday or Friday could have returned home on Saturday before the storm struck that border that night.

But it's not an easy call. The NHC's now famous "projection

cone" increases in width as the distance (and time) from the storm center increases. It is an excellent visual depiction of the range of error in the prediction. Introduced in the mid-1990s, the cone has become narrower and narrower, as the forecasters have become more accurate and confident, but the margin of error three days out is still 250 miles in each direction. One day out—twenty-four hours—it is 85 miles in each direction. Many people are surprised to learn—*I* was surprised to learn, from *Divine Wind*—that the best our science can ever achieve will be an error of 90 miles at three days, 30 miles at one day. The projection cone will never be a straight line.

This rude fact of life about chaos as it relates to weather was first demonstrated—by accident—in 1961, by Edward Lorenz, a professor of meteorology at MIT. The story has been told time and again, but its import is so integral to weather forecasting of any sort that I'll tell it again now, in the short form. Holding one set of answers to a problem in his hand, Lorenz decided to run the same calculations again and extend them further in time. In order to make this new job more manageable for the painfully slow vacuum tube computers of that era, he rounded off the variables to the thousandth place. That is, 2.378258 became 2.378, thus introducing an error of, at most, one in a thousand. This was not much, and in some systems such small differences in the initial state would yield small differences in the end state. In Lorenz's calculations, however, they yielded huge differences in the "prediction," so huge that he initially thought there must have been some mistake. But there was no mistake. He had stumbled across a bedrock principle of chaos theory. More broadly, the theory limits the most accurate imaginable weather predictions for a given locale to two weeks out. Better than that is now believed mathematically impossible.

Whatever the cost of narrowing the forecasting error for hurricanes to the theoretically achievable threshold, the expenditure will be worth it. Short distances mean everything with a hurricane. The truly devastating winds and surge zone of even the most dangerous storms is rarely more than one hundred miles across, and often much less than that. A shift in landfall of thirty miles will make a world of difference regarding not only the damage incurred, but the evacuation necessary to flee that damage. The theoretically achievable accuracy of the computer models would have made most of the evacuations for Georges, Floyd, Ivan, and Rita—to name just four noteworthy examples—completely unnecessary.

Beyond the economic disruption, another problem with unnecessary evacuations is that some residents eventually get jaded and don't leave the next time. Research shows that evacuating types usually do evacuate, but after two years of almost unending watches and warnings and evacuations, the citizens of Key West, Florida, were not in the mood with Hurricane Wilma in October 2005. Many ignored the mandatory evacuation order for the Florida Keys. Regarding Katrina, Governor Blanco would refer to the phenomenon when testifying before the congressional committee investigating the response: "You put your four kids in the car and you're sitting in traffic and they're screaming and nothing happens and you go home and say, 'I'm not doing this again. This is crazy.'" Overall, however, the problem with Katrina and New Orleans is more complex than for perhaps any other locale in this country. Our LSU surveys have always showed that 30 percent of the residents of New Orleans will not evacuate, the majority because they are "low mobility," in emergency preparedness jargon—that is, they have no ready access to a car. Others shared the jaded view of the carriage driver in the French Quarter who told a reporter, "They've been singing this song for thirty years, and what? Nothing." And perhaps a number of residents share the fatalism of the Druid priest John Martin, a resident of the French Quarter (naturally), owner of four snakes (including the giant Burmese python Eugene, somehow missing as Katrina approached), who announced to any reporter who would listen over the weekend, "I don't believe you're going to go until God takes you. I've lived a good, full life and I'm not worried about it. You've got to take life as it comes."

LSU sociologist Jeanne Hurlbert, who teamed with her husband, professor of sociology Jack Beggs, to conduct our evacuation surveys in 2003 and 2004, quipped, "There's a reason New Orleans is famous for the drink named the Hurricane. The culture here is 'We don't evacuate.'" I think that attitude has softened somewhat in recent years. The "Washing Away" series in the *Times-Picayune* had a powerful impact on many people. I like to believe the drumbeat of warnings issued by myself and many others has also had an impact. Certainly the compliance for Ivan in 2004 was much better than it had been for Georges six years earlier, when only one in three residents left town. Then in 2004 we had the three major storms in Florida, followed by the unbelievably catastrophic tsunami in Asia in December, which certainly had an impact on public attitudes. From that day on I often invoked the tsunami to dramatize the damage that could result from a major storm in New Orleans. I always invoked it in local TV interviews. When we were trying to get people to heed Mayor Nagin's plea for a mandatory evacuation (which finally came on Sunday morning), I said in a few interviews that "this could be our tsunami." An exaggeration, definitely, because well over two hundred thousand lives were lost in the tsunami, a number that would probably never be the case in New Orleans, but that's okay. The tsunami was a fresh image. I got a handful of e-mails criticizing the analogy, but they were water off this duck's back, because we needed to scare as many people as possible to get out of the city. Desperate times call for desperate measures. I wasn't the only talking head to invoke the tsunami, and I'll bet all of us combined received literally thousands of e-mails thanking us for doing our best to clear everyone out.

In the end, according to the calculations of the Department of Transportation and Development, 430,000 cars left metropolitan New Orleans and coastal areas to the south during the evacuation. Figuring (on the basis of previous research) 2.5 persons per car, the total is 1 million evacuees, 75 percent of the total population. Truly a remarkable achievement. Our conclusion that 30 percent would never leave, under any circumstances, was off by 5 percent. Thank God.

Still, more than three hundred thousand people remained in New Orleans and vicinity, for good reasons or bad reasons or no reason at all. What was the plan to deal with these folks? Time and time again over the following weeks, I would hear individuals say that everyone who remained behind had sufficient warning, could have found a way out, and therefore pretty much got what they deserved. I guess such hard-hearted observers thought of the people who didn't leave as something like heavy smokers: Okay, it's a free world (in some places), but don't come crying to me or my insurance company when you get lung cancer. Forget the mean-spiritedness. As a matter of public policy, such attitudes are irrelevant. So let's get such cheap thinking out of the way right now. No government—no nation—can sit by and watch tens or hundreds of thousands of people drown or otherwise die, even if it is the result of their own bad decision. As a matter of public policy, every official in the state of Louisiana knew that over three hundred thousand people would remain in the New Orleans metro area. Again, what was the plan to deal with them?

Years ago in South Africa I saw an old Charlton Heston movie, *The Omega Man*, some kind of doomsday apocalypse narrative featuring an empty city, and I distinctly remember wondering at the time, "How did they get everybody out?" This was before digital manipulation. It looked like a real city that really was empty. Over the following weeks in New Orleans I would think of that scene often and wonder how they got all of those people out. Maybe FEMA should have hired Hollywood.

At 9:58 P.M. Saturday ADCIRC produced the results of the surge model based on advisory #18, which had been issued five hours earlier. Run #16 in the early afternoon had looked ominous, #17 equally ominous, but here was the worst news in the vivid colors used for the models. New Orleans would flood. On Monday a large part of the city east of Industrial Canal would be under some water, as well as other sections to the west of it. All of this flooding would result from the overtopping of levees, *not* from any breaches or failures. Clearly, the models can't account for such failures, because we can't know where they would be. Correlate this latest prediction with the fact that the National Hurricane Center had issued a warning at 6:00 P.M. that the city had a 45 percent chance of receiving a direct strike from a Cat 4 or Cat 5 Katrina, and, well, this was bad. Evacuate now! Plus, the prediction could get even worse. The hurricane was still thirty-six hours away, and still strengthening. I said to Marc Levitan, “This is the one we’ve been working on. This is the big one. Let’s hope it goes east, for New Orleans’s sake.” I think that’s just about an exact quote.

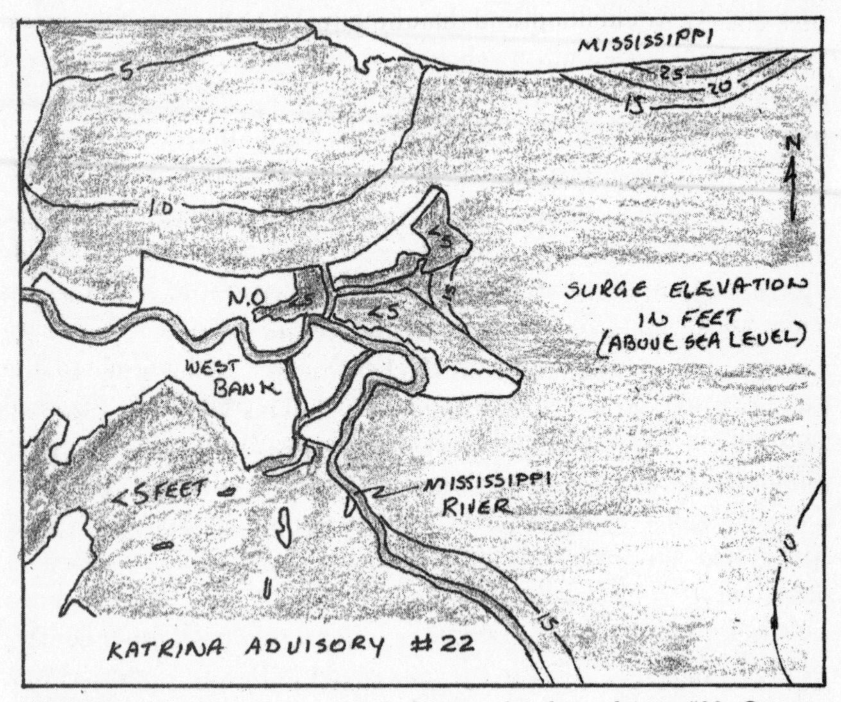

ADCIRC surge model with predicted elevations based on advisory #22. Contours every five feet; elevations above mean sea level.

It was a strange moment for all of us in the local “hurricane community.” In Ivan the year before, the surge models had never predicted the flooding of New Orleans. For Georges in 1998 we didn’t have nearly as good a product. Here, though, in real time, was a prediction we had confidence in. Most of us were jammed in Mashriqui’s cubicle of an office when the pivotal graphic rolled out of the printer. We fell silent for a few moments. We’d been warning about this for years, but to actually “see” the flood in the ADCIRC graphic was truly shocking. Everything we had been talking about and studying for years was about to unfold before our eyes. I got a strong chill all over, and a real tightness of breath. (As an asthmatic, I'm always aware of this kind of thing.) Then we sprang into action. Someone may have said, "Let's get this out," but it wasn't really necessary. All of us knew what to do. I phoned Mark Schleifstein at the *Times-Picayune.* He was waiting for the outputs. We told everyone that the city would flood, and that the actual flooding would be even worse than our surge model had predicted, because some levees were believed to be below their announced heights and a wild wave field would develop on lakes Pontchartrain and Borgne.

Run #18 was projected onto the big screen at the state's EOC at 11:00 P.M. In our briefing we advised everyone present—FEMA, the National Guard, all state and local agencies—that the actual flooding would be worse. We reminded them that the levees along Industrial Canal had been breached during Betsy in 1965. (The reception was quite different than when Marc had presented the model based on advisory #16 earlier in the afternoon, that had showed ominously high surge levels. A senior Louisiana emergency official had seemed uninterested in that output. Subsequently, other LSU staffers complained that the new upper echelon of the state emergency agency was not utilizing the science, unlike the management the year before during Ivan.) If we had such flat-out failures this time, the flooding would be much, much worse. On a conference call from Miami, Max Mayfield, director of the National Hurricane Center, briefed Governors Blanco and Barbour and New Orleans Mayor Nagin. Get everyone out, he said.

Every media outlet was all hurricane, all the time, and they all had the news. No one paying the least bit of attention could have avoided it. No one. So when Michael Brown said days later that on Saturday it was "my belief . . . we had a standard hurricane coming in here, that we could move in immediately on Monday and start doing our kind of emergency response effort," I was livid, because we had called it right and called it early. Our warnings went out at least thirty-two hours before Katrina made landfall. That was ample time for FEMA to act. Get those buses rolling—no matter where you have to get them. This is the United States of America, not some desperately poor country with no resources. (The New Orleans firefighters deserve a word of praise at this time. Chief Gary Savelle gave his crews copies of the ADCIRC graphic I had e-mailed, and they started telling folks in the areas that would flood to *leave*. They acted in a responsible manner.)

Late Saturday night I worked with Mark Schleifstein on his story for the Sunday edition of the *Times-Picayune*. We both thought this was tremendously important, because Sunday would still give people enough time to leave—at least those with cars or access to cars. The following morning, the whole package was headlined KATRINA TAKES AIM, with three subheads above the fold on the front page: "An Extreme Storm," "Get On the Road," and "Wall of Water." The graphic showed in detail how the flooding would occur. As it turned out, many traditional nonevacuators did change their minds. They fled on Sunday, maybe with the naive notion that they'd get a motel in Baton Rouge, only to end up in Arkansas instead, but they did make it out. On the other hand, the parties on Bourbon Street were in full swing all night long. So I understand. But if just one person was saved by all our hard work—just one child—then it was worth it.

THREE

LAUGHED OUT OF THE ROOM

While the oblivious partied in the French Quarter on Saturday night, some of us tossed and turned. We were quite sure the fun was just about over, not to resume for quite awhile. In fact, Max Mayfield of the National Hurricane Center said on Sunday, "New Orleans is never going to be the same." Wow. I know Max to be a levelheaded guy and not prone to exaggeration. Then again, he can read a map and the data with the best of them, and he had stated on the air on Saturday that he'd never seen conditions so ripe for hurricane strengthening. It happened. On Saturday night Katrina exploded and officially reached Cat 4 status—sustained winds of 130 mph—at 1:00 A.M. Sunday morning. At 8:00 A.M., advisory #22 announced a Cat 5 storm headed for the Gulf Coast with sustained winds approaching 175 mph and higher gusts. The 10:00 A.M. advisory added "potentially catastrophic" to the description. Such a storm would flood New Orleans to the eaves *and* tear it to shreds. Likewise for all surrounding communities. An old saw in the disaster field holds that water damage starts from the bottom up, wind damage from the top down. If they meet in the middle, it's all over for that building. Based on these latest bulletins, such was the prospect with Katrina. Our LSU wind-modeling team estimated two hundred thousand severely damaged structures in a direct hit by such a storm. (This model correlates predicted winds with a modified FEMA database of all structures in given zip codes.) Wind engineer Marc Levitan said on Saturday, "If everyone doesn't get out, with the wind and the water, there is strong potential for mass casualties."

Over the weekend several factors played right into Katrina's hands. For starters, the Gulf of Mexico was even warmer than usual, 85 degrees or higher, mainly because of the insistently hot weather along the coast for weeks, perhaps also because we had had very little rain inland, therefore less relatively cool freshwater runoff. There were no big shearing factors on the horizon for Katrina, and a high-pressure ridge in the upper atmosphere was serving nicely to vent the heat from the highest reaches of the storm. (One of the incredible graphics put together by Kerry Emanuel for *Divine Wind* depicts the rising heat in a hurricane, with the hottest zone at the top of the eye—heat that is whipped away in a clockwise direction. In the strongest storms this venting process is achieved efficiently, thereby sucking up more heat from below: more wind—more strength.) Finally, on Saturday the storm had undergone an eye-wall replacement cycle, in which the innermost wall of high winds and clouds breaks down and is replaced by another wall. This cycle can lower the winds a bit—only to be followed, often enough, by an increase. Max Mayfield and his team knew that the replacement cycle during the day Saturday could be followed by strengthening that night and Sunday. So it was. The only good news was that the burst of energy that had taken the storm to devastating Cat 5 status would probably not hold up. Katrina could lose a little strength late Sunday while approaching the coast, but it probably wouldn't matter all that much. Whether she struck as a Cat 4 or 5, the flooding would be about the same, according to our surge models.

The big question was always the track, with the eventual landfall determined primarily by the influence of two high pressures and one low pressure. For the moment, a high pressure over the coast north of Katrina was steering the storm to the west, the same kind of action by which the Bermuda High steers many Atlantic storms to the west. But a turn more to the north was expected, due to the influence of, in the words of the NHC, "a weakness in the ridge associated with a large midlatitude low-pressure system over the northern United States and southern Canada." So our violent visitor would slowly and almost inexorably turn to the northwest, then the north. But how much? On Saturday morning the zone of uncertainty had stretched from the Louisiana/Texas border to Pensacola, Florida, with the main track right over New Orleans. That was quite a spread. By that afternoon the track had shifted slightly to the east, and the cone had narrowed slightly, but the storm could still go west of New Orleans. When we restarted our storm-surge modeling early Sunday morning with advisory #22 the storm was still on virtually the same track as with #18, the advisory from the night before that had first showed the flooding. The strengthening overnight offset the slight move to the east. The new flood forecast remained about the same. A slightly more definitive move to the east, however, would be very good news for New Orleans, just as a shift to the west would be very bad news. The absolute, 100 percent worst case would be for Katrina to make landfall at Grand Isle, almost due south of the city, and proceed to the north-northwest on a track taking her eye just west of the city. If this happened, the ADCIRC storm-surge output would look considerably different. We'd have something equivalent to the storm we developed for FEMA's Hurricane Pam exercise—a completely flooded city, with water over the tops of all the levees. Katrina certainly had the power to do this, but only if she tracked in the western part of the projection cone and slowed down just a touch. Forward speed is very important for the surge equations. The slower the storm, the longer the time for the surge to build and build. We always fear a slow-moving storm.

What did we know? What did we predict? What would happen?! Most of the reporters and TV producers were earnestly trying to get the best information they could. They were very polite and keenly interested in our data and opinions. We sent them to our various Web pages. A few requests were peremptory, as though they were doing us a favor. Some were ingratiating, a few pretty clueless. I—and I assume everyone else on the staff—got e-mail requests asking for thoughtful analyses of the long-term consequences of this storm as it would affect this or that specific issue. "Of course, I know you are busy today," would be the obligatory caveat. Well, yes. All of us ended up, over the next few weeks, with literally thousands of unanswered—unopened—e-mails and messages on our phones.

At 10:00 A.M. Sunday I joined a conference call with Dr. Jim Diaz of LSU's medical school and with Martin Kalis and a large team at the Centers for Disease Control in Atlanta. Within half an hour of our posting of surge-model #18 the previous evening, Martin had called to set up this meeting to discuss the public health problems that would follow hard on the heels of Katrina's imminent arrival. Then Martin sent us an e-mail (time-stamped 5:05 A.M.) with a revised list of the potential health impacts from the expected flooding, and it was a long list, all entries previously discussed, including everything from massive petrochemical spills during the storm to mildew issues for years to follow. Fire ants, too. Earlier in the week, knowing that I often mention these insects as a hazard during floods, Martin had forwarded a photograph of a fire ant colony floating on top of floodwater. I knew all about that, in part because a pasture rich with fire ants behind my home had flooded with the passage of tropical storm Alison. Donning my rubber boots, I searched for and found many clumps of ants about the size of tennis balls. Does the writhing, rotating action of the ball assure that a given ant will not be underwater long enough to drown? That's a pretty good hypothesis. Anyway, it was quite fascinating, but then I got into my scientist mode and experimented with several different insect killers. The winner was definitely the liquid jet-type hornet and wasp spray.

At some point during the Katrina emergency, a friend sent me a blog associated with the *Columbia Journalism Review* suggesting I must be some kind of nut for even mentioning the fire-ant hazard. I deny the accusation. The ants do interest me, but the problem is real regardless. The average yard in Louisiana has two fire-ant mounds, and during a flood these floating colonies will tend to end up where the humans do—on trees, for example. Any such encounter will turn out very badly for the people. In the South there have been instances in which an attack of fire ants caused people to release their holds on trees and drown in the flowing waters below. During Katrina, as it turned out, the number of fire-ant problems was minimal, probably because more people ended up on rooftops, not trees, for survival. However, LSU Agricultural Center entomology associate Patricia Beckley has documented that some people wading through the waters surrounding the Superdome encountered masses of floating fire ants. Elsewhere, pesticides and thin oil and gas sheens floating on the surface of the waters may have killed many of the colonies.

The CDC's two-page list of health impacts included West Nile virus, rabies, waterborne gastrointestinal diseases, burns, pulmonary irritations—the usual suspects, and more—and it was the job of both our center at LSU and the CDC to be prepared to provide relief officials and policymakers with the best information regarding all of them. The CDC provides medical support teams to any state requesting their help, and it also has rapid-assessment teams, a critical part of any response to a major disaster. They can assure that the correct medical supplies, health resources, and manpower are available as soon as possible. Because of our research we could give them a good heads-up on what to expect after Katrina.

I was not surprised by the CDC's prompt and efficient response to the telltale surge model. They're on the ball. Compared to some of the FEMA officials with whom we had dealt in the past few years, they are disciplined, scientific, fact-based, and results oriented. We had spent a day with them in Atlanta in October 2004, then later that year a group had come down to LSU, and we all went down to Charity Hospital in New Orleans. We would have liked to work more closely with them, but Bush administration cuts to their budget were so severe that Martin Kalis couldn't get the travel funds to come to our advisory board meeting in 2005.

Sunday morning, Mayor Nagin at City Hall finally ordered the mandatory evacuation, the first in the city's history, with Governor Blanco by his side. Finally. I know that "mandatory" is just a word, and it doesn't really mean that the cops are ordering everyone out at gunpoint, but it's a scarier word than "voluntary" and should have been invoked on Saturday, legalities be damned. We were now less than twenty-four hours before landfall. "We are facing the storm that most of us have feared," the mayor said. "This is a once-in-a-lifetime event. It most likely will topple our levee system." He had obviously seen our surge-model outputs. I assume he had seen that Sunday morning edition of the *Times-Picayune*. What a sad moment for this former businessman, who just the week before had been gloating about Donald Trump's impending visit to announce his new condo hotel on Poydras Street. Now the first-term mayor was likely to preside over an incredible catastrophe that would find most of Poydras Street under water. This was one of the announcements at which citizens were advised to have an ax handy if they refused to leave, because they would need it to chop their way out of the attic. I heard that some of the crews driving around town announcing the mandatory evacuation through bullhorns added dramatically, "Run for your lives." People did. Everyone had seen the tsunami footage. At one point on Sunday, eighteen thousand cars an hour were leaving the city. Slow going, but these last-minute leavers did get out.

Hotels become last-resort zones of "vertical evacuation." Although the hotel managements had always resisted any such official designation, they knew that they would fill up to the brim, and they did. I wonder how many of the people who stayed in the hotels know that the wind increases the higher one proceeds "up" the hurricane. Today's modern skyscrapers, including the hotels, are designed to withstand blasts of 120 mph, and I'm sure they would do so, but I can't imagine a more terrifying experience than huddling in the hallway on the tenth floor as the shrieking winds of a Cat 4 or Cat 5 hurricane shatter every window of my swaying hotel. I think I'd run for the hills instead. (Some high-rise buildings use pea-sized gravel as a roof covering. In a storm, this gravel becomes lethal window-smashing missiles. It is not rare to see, after a storm, skyscrapers with most of the windows on one side smashed, up to about 150 feet above the ground.) Most modern buildings have few strong, solid interior walls; office spaces are created with modular units or flimsy partitions. Once the windows have been blown out, these "walls" may be the next to go. Even if they don't blow away, the wind flow may create tunnels filled with blowing debris. Not a safe place with winds of even 60 mph. Imagine 120 mph, a couple of hundred feet in the air. Vertical evacuations are really a matter of absolute last resort.

The Superdome had been opened as a "special needs" shelter on Sunday at 8:00 A.M. for those residents with a good reason not to evacuate. Theoretically, people were even supposed to call a special number to see if they qualified for admission—a plan that never had a prayer. Colonel Terry Ebbert, the city's director of Homeland Security, explained that the infirm would be directed to one side of the building, where "we have some water." Those seeking shelter were advised to bring enough of their own food and water for a few days, but they should not plan on staying long. (And they should then proceed . . . where?) By noon Sunday, thousands of residents had lined up. Few had their own food and water; none had anywhere else to go or any way to get there. CNN reported that nine thousand men, women, children, and babies spent Sunday night at the Superdome. FEMA said fifteen thousand meals were on hand, but I also read that forty-one thousand meals were gone within the day. As with many of the numbers all of us read and heard over the coming days and weeks, I have no idea which ones are correct. I don't think anyone does. Contradictions abound. Someday, a group of academics or reporters, or both, with the time and resources will try to find out. This is going to be quite a challenge, and I'm not sure any such postmortem will clear up all of the questions. There was some food and water at the Superdome. We know this. But not enough. We also know this.

In their Sunday edition, *Times-Picayune* editors published a list of twelve locations from which people would be transported to the Superdome. The streetcars on St. Charles Street would also be operating for as long as they could. One terminus was less than a mile from the Superdome, and this system did bring a few thousand people to the big building. Others walked. Chester Wilmot, my colleague at LSU, reports that the city had 550 city buses and hundreds of other buses at its disposal, but no plans to use them in an evacuation. The goal should have been "transportation redundancies," in Wilmot's phrase. In 2001 the LSU Hurricane Center did a review of the effectiveness of buses, but pointed out that the 550 city buses plus "hundreds" of others, while it sounds like a lot, would not have been an effective means of getting the 127,000 persons without vehicles out of New Orleans. As it turned out, the buses sat in the various parking lots—flooded, many of them, as of the following day. To my knowledge, *not one group* of the infirm or elderly was evacuated by an "official bus" prior to the storm.

Two weeks after Katrina hit, on *Meet the Press* Tim Russert challenged Mayor Nagin about the flooded buses. The mayor said there had been no drivers. Moreover, he asked rhetorically, where would the buses have gone? This was a good excuse! Have a plan for the drivers. Have a plan to provide safe shelter for them after the work is done. Have a set of destinations prepared. On his show, Russert, like all of the reporters by that point, was in no mood for Nagin or anyone else's rope-a-dope. He quoted New Orleans's own emergency plan: "Conduct of an actual evacuation will be the responsibility of the mayor." Nagin replied that the overall plan was "getting people to higher ground, getting them to safety . . . and then depending upon our state and federal officials to move them out of harm's way after the storm has hit." The cavalry would come. That was the hope and expectation. At other times Nagin had referred to this strategy as Plan B: "The president, I'm sure, is going to send us what we need."

In short, the city's plan amounted to a "good samaritan" response, in the phrase of Brian Wolshon, transport engineer and a fellow researcher at our Hurricane Public Health Center. This was not good enough. In 2004, the American Red Cross had started looking into a program called Brother's Keeper, whereby religious organizations would ferry some of the immobile in New Orleans to safe church shelters outside of New Orleans. Not good enough. Some private citizens' group also had some plans. Not good enough. As wonderful as such gestures were, they were still just *gestures*. They could never evacuate more than mere thousands, when the problem was tens of thousands. The nursing home trade group for Louisiana concluded after the flood that at least two thirds of the city's fifty-three nursing homes were not evacuated, with tragic results.

In mid-February 2006, the Senate Homeland Security and Governmental Affairs Committee revealed that the State Department of

Transportation and Development had in April 2005 been tasked with developing plans and procedures to mobilize transportation to support emergency evacuations of at-risk populations. The state DOTD does not have buses or drivers, and so did not move on this responsibility. Secretary Johnny Bradberry was severely chastised by senators Susan Collins and Joe Lieberman, given the number of deaths during Katrina at hospitals and nursing homes. All reports are that Bradberry took his medicine without trying to pass the buck, something that seems to have been the hallmark of others this committee found fault with.

Tim Russert challenged Mayor Nagin on the public service announcement he had cut in July about evacuating the city. In Russert's phrase, the mayor had warned citizens that they were "on their own." An exaggeration, but several times the mayor had expressed doubt about a successful evacuation of his city. In July, he had said that a Cat 5 would be an "easy sell," in terms of evacuation. For anything less, he added, "The community says, 'We might ride this out.'" This belief about the Cat 5 factor was optimistic, of course. He surely knew of our widely disseminated conclusion that about 30 percent of the residents would not leave the city regardless. He must have believed it. As it turned out, as few as 25 percent stayed behind, which was great, but where was the plan for establishing safe higher ground for them? And what about the elderly and the disabled?

Russert informed Nagin that New Orleans had received $18 million since 2002 to plan for just such emergencies. Where had the money gone? "Levee protection and the coordination of getting people to safety," the mayor replied. What could he say? I actually felt sorry for the man. It's easy to criticize in hindsight, easy to criticize from the sidelines—and it seems to be easy for a good interviewer to make almost anyone look bad on television (as FEMA's Michael Brown would whine by way of self-defense during a remarkable interview on PBS's *Frontline* in November). However, the fact remains that the government does not have the luxury of the private citizen who dismisses those who stay behind as equivalent to smokers who get what they deserve. That's neither a practical nor a morally responsible public policy.

Testifying before Congress in December, both Governor Blanco and Mayor Nagin were accused of waiting too long to issue a mandatory evacuation order for New Orleans. The governor pointed to the extraordinarily successful evacuation: 92 percent of the residents of southeast Louisiana. (This is a very high number. Our best figure, as reported, is a still excellent 80 percent.) Nagin admitted that he should have handled the bus problem better. Indeed. The photographs of the flooded buses—the yellow roofs neatly lined up in the black water—will remain forever as one of the lasting images of Katrina in New Orleans, an iconic depiction of the horrible planning.

Over the following days and weeks we would hear a great deal about the legal responsibility for what happened in New Orleans. The simple truth is that under the Stafford Act (§401 and §501) of October 2000, once President Bush declared a national emergency on Saturday, August 27 (retroactive to the day before), the federal government was in charge. To my knowledge, the first reporter in the national media to get this right, postdebacle, was NPR's Ira Glass, on the September 9 edition of *This American Life*. Sure, Glass said, "there are plenty of things that state and local government did to screw things up," but the declaration of national emergency made all that moot, in terms of responsibility for what happened next. He was interviewing William Nicholson, author of the book *Emergency Response and Emergency Management Law*.

Of course, you have to plan ahead for such an eventuality, don't you? It would have been a daunting challenge, to say the least, for FEMA to have jumped in over the weekend to organize an extensive evacuation of the immobile without such prior planning. But FEMA knew the numbers. It knew the repercussions of heavy flooding. It knew what the city had and had not done to prepare. All of us in the disaster field did. The previous year city officials had been caught off guard and had opened up an unprepared Superdome just twelve hours before Hurricane Ivan made landfall to the east. Only a few thousand people showed up, and the city didn't flood (and was never expected to), but the absence of a workable plan was obvious. A year later it was already obvious again on Sunday.

It was FEMA's responsibility—by law—to serve as the cavalry—

and not just after the fact. I had walked away from the Hurricane Pam exercise one year earlier (to be discussed in some detail in Chapter 7) with the conviction that FEMA would assist in the evacuation of the city prior to any major storm. After Katrina struck and everything broke down, I asked Walter Maestri, the highly respected director of Emergency Preparedness for Jefferson Parish, whether this was also his understanding. "Absolutely," he replied.

Imagine. Almost all the suffering and displacement that followed Katrina could have been alleviated with just a few evacuee camps—tent cities—within fifty miles of New Orleans and an efficient system for getting people there. I had seen for myself how well and quickly these camps can be organized. In July 2003, my colleague Kate Streva and I had taken an intensive, ten-day course in refugee camps sponsored by Merlin, a nonprofit organization based in the United Kingdom. About thirty of us from many different countries, some with a wealth of experience, attended our classes in the peaceful English countryside of Sussex. The experience was quite an eye-opener for the two (and only two) Americans. Kate and I had known that a major hurricane strike on southeastern Louisiana would leave hundreds of thousands homeless and jobless, without transportation or schools, for months, if not years. Now we knew the best possible solution: evacuee camps, or whatever name you choose to call them. (Officially, a refugee refers to an individual displaced from another country. An individual displaced within his or her own country is an "internally displaced person," or IDP. During the Katrina emergency, I tried to use "IDP," but it never took. Instead, refugee was used for a few days before "evacuee" became the accepted description.) We may shrink in horror from the idea, FEMA may refuse to entertain it, but these tent cities are the easiest way to provide large numbers of evacuees with the required number of medical clinics (one per five thousand), latrines, cooking tents, and so on. They can be erected very quickly—within hours.

The FEMA folks not only refused to entertain the idea. They basically laughed me out of the room when I broached the subject during the Hurricane Pam exercise. They didn't want to hear about the army's proficiency in erecting such cities. (Just look at our efforts in

Bosnia and Herzegovina. And Iraq. Everyone has seen footage of our troops and has a mental image of these tent setups, with wooden floors and electric lights and cooling fans and privacy.) I suggested prepositioning tents and supplies, as well as signing leases with major landowners north of Lake Pontchartrain, in St. Tammany and Tangipahoa parishes. The laughter got louder. The woman said, "Americans don't live in tents!"

Undeterred, I tried to push this concept during the Katrina tragedy, in interview after interview. It was too late, I realize, but I wanted people to understand the alternative to what was going on. We could have had comfortable facilities ready to go, complete with large tents serving as cafeterias, and as halls for entertainment and lectures on dealing with all the paperwork problems that would follow. We would have had a ready workforce of residents, once the cleanup started. Instead, we ended up with the local workforce scattered in one thousand different cities across the country, some still in hotels four months after the storm, some still in the makeshift tents they chose over the other options presented. Evacuees were arriving in Baton Rouge without any medical support. Diabetics, heart patients, and asthmatics were on their own, initially. After the flooding began, many had severe staph infections from being in the water so long. Others waited and waited, some for seven days, before they were bused all over the country with no real medical surveillance until they arrived at their destination. The Astrodome in Houston, where most from the Superdome were taken, filled to overflowing and had to close its doors. Some evacuees were put on planes and not told where they were going. (I wonder if the pilots knew when they took off. I wouldn't bet on it.)

All because "Americans don't live in tents." FEMA refused to even think about the most obvious option for dealing with the inevitable numbers of evacuees following the event that the agency itself had long recognized as one of the three most likely and dangerous disasters for the country, along with a terrorist attack in New York and another major earthquake on the West Coast. Americans need to understand that their government is totally unprepared for major natural disasters, let alone the terrorist's dirty bomb or biological/

chemical attack. Don't kid yourself. Very little, if anything, has changed since September 11, 2001. Domestically, that is. On the West Coast, in New York City, in Washington, D.C., in Chicago, pay heed.

On Sunday in Baton Rouge I didn't know all the details about the unused buses or the brewing travesty at the Superdome, but the networks were already carrying footage of families lining up outside the building as every public official in Louisiana, Mississippi, and Alabama issued new warnings and orders. In Texas, President Bush held a special press conference vowing that the federal government "would do everything in our power to help the people and communities affected by" Katrina. Max Mayfield briefed him by videoconference. FEMA spokesman David Passey said, "This looks like it could be worse than any natural disaster in the U.S.—ever." At LSU our phones were ringing off the wall, and Marc Levitan, Paul Kemp, and all of us took every opportunity to invoke keywords like "worst-case scenario" and "catastrophe" and "thousands possibly dead" and "EVACUATE." Someone quipped that the Big Easy was now the Big Queasy.

At 10:11 A.M. Sunday the National Weather Service office in Slidell released a remarkable warning. At the time, I didn't recall another like it, and I've since learned that nothing like it had ever been issued. "Most of the area will be uninhabitable for weeks . . . perhaps longer," this bulletin stated. "At least one-half of well-constructed homes will have roof and wall failure. The majority of industrial buildings will become non-functional. Airborne debris will be widespread. Power outages will last for weeks. Water shortages will make human suffering incredible by modern standards."

At 2:56 P.M. the surge warriors released ADCIRC model run #22, which showed a little more flooding than the previous ones, but in the same areas. The cone of uncertainty was getting tighter and tighter, and the track had not changed. The eye would probably move right over the east side of New Orleans. "Will the levees fail?" That was always on our minds. From much experience, I knew there would be a wicked wave field on all the surrounding lakes, and especially on Lake Borgne and Lake Pontchartrain. These waves could well erode the earthen levee systems (which proved so true in St. Bernard Parish).

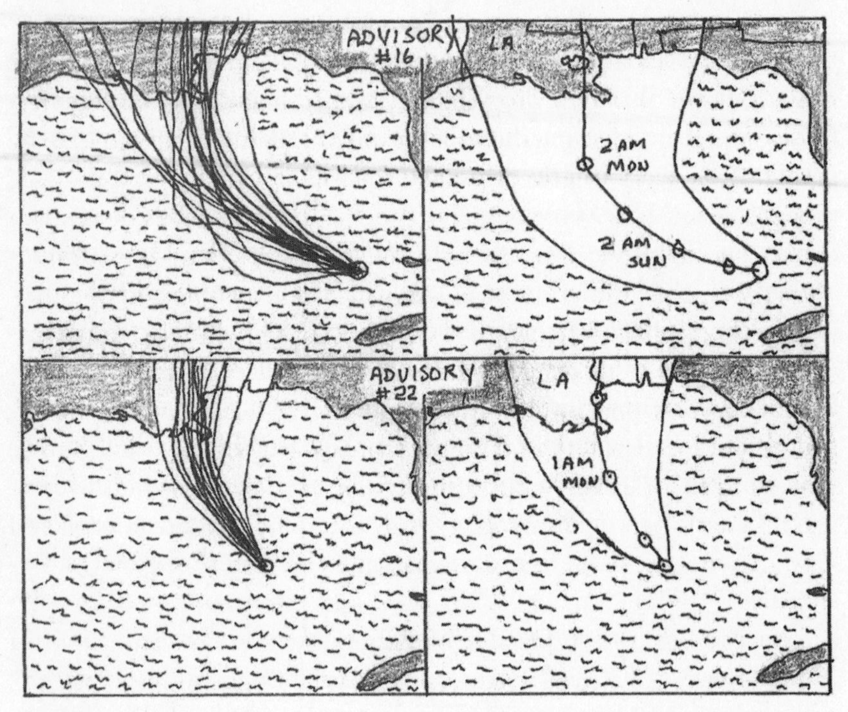

Spaghetti plots and cones of uncertainty for Hurricane Katrina advisories #16 (August 27, 2005) and #22 (August 28, 2005).

Since the latest model showed no significant change in the flooding advisory, I told every one to take a quick break, get their homes and families sorted out, and reassemble around 10:00 P.M. I suddenly caught the fear bug and decided I'd better beef up the defenses on the *Maggie*, so Lorie and I raced out and added two more anchors and stern lines. As we departed the anchorage I told her we might well lose the boat this time. In a couple of days we might be looking at the mast alone, a heartbreaking prospect. In December 1999, we had hauled the *Maggie* and for the next two years spent every other weekend on a stem-to-stern refurbishment at the boatyard in

Madisonville. That was a family project; our two daughters, then quite young, helped out with the odd small painting job and making sandwiches for lunch. Now I tend to relate such wonderful memories to those of the flooded families in New Orleans.

At 10:00 P.M. Sunday, Mashriqui sent out an e-mail stating that run #25 showed *less* flooding in New Orleans. Katrina's predicted track had moved a bit east and her winds were down slightly. Very important but also very dicey news. Would people lower their defenses just a little, even though the potential for flooding in the greater New Orleans area remained the same? I e-mailed the latest prediction to Mark Schleifstein at the *Times-Picayune*, and I immediately phoned Luigi Romolo, who was running the LSU briefings at the EOC, and told him *not* to lower any of his warnings. We definitely did not need to be undermining people's vigilance or changing a very last minute decision to leave. In an e-mail to Luigi I added a note—for the first time—that this more optimistic latest prediction from the surge model assumed that the levees held. I reiterated that a breached levee on the Industrial Canal caused much of the worst flooding during Hurricane Betsy in '65. At about this time, on CNN Paul Kemp answered a suggestion that the reduced wind speed was good by saying, "Well, that's not going to be a lot of solace for people in New Orleans, because that storm will also flood New Orleans. And what we're concerned about is getting people out of there." Referring to wind speeds, Paul said, "Ten miles an hour one way or the other is not going to make a big difference. This is a killer track." That became the new mantra. Marc Levitan showed me the latest results from the wind-damage model, which did not look good. In some areas south and east of New Orleans, 80 percent of homes would be badly damaged.

As dawn broke through the windy, rainy skies on Monday morning, with the eye of Hurricane Katrina just reaching the farthest marshes of St. Plaquemines Parish 125 miles away, I found the CNN truck outside the LSU dormitory where I had lived as a first-year graduate student in 1977. Twenty-eight years later, there I was talking with Miles O'Brien about the major hurricane barreling down on us—the "product," in a way, of my life's work. (Somehow

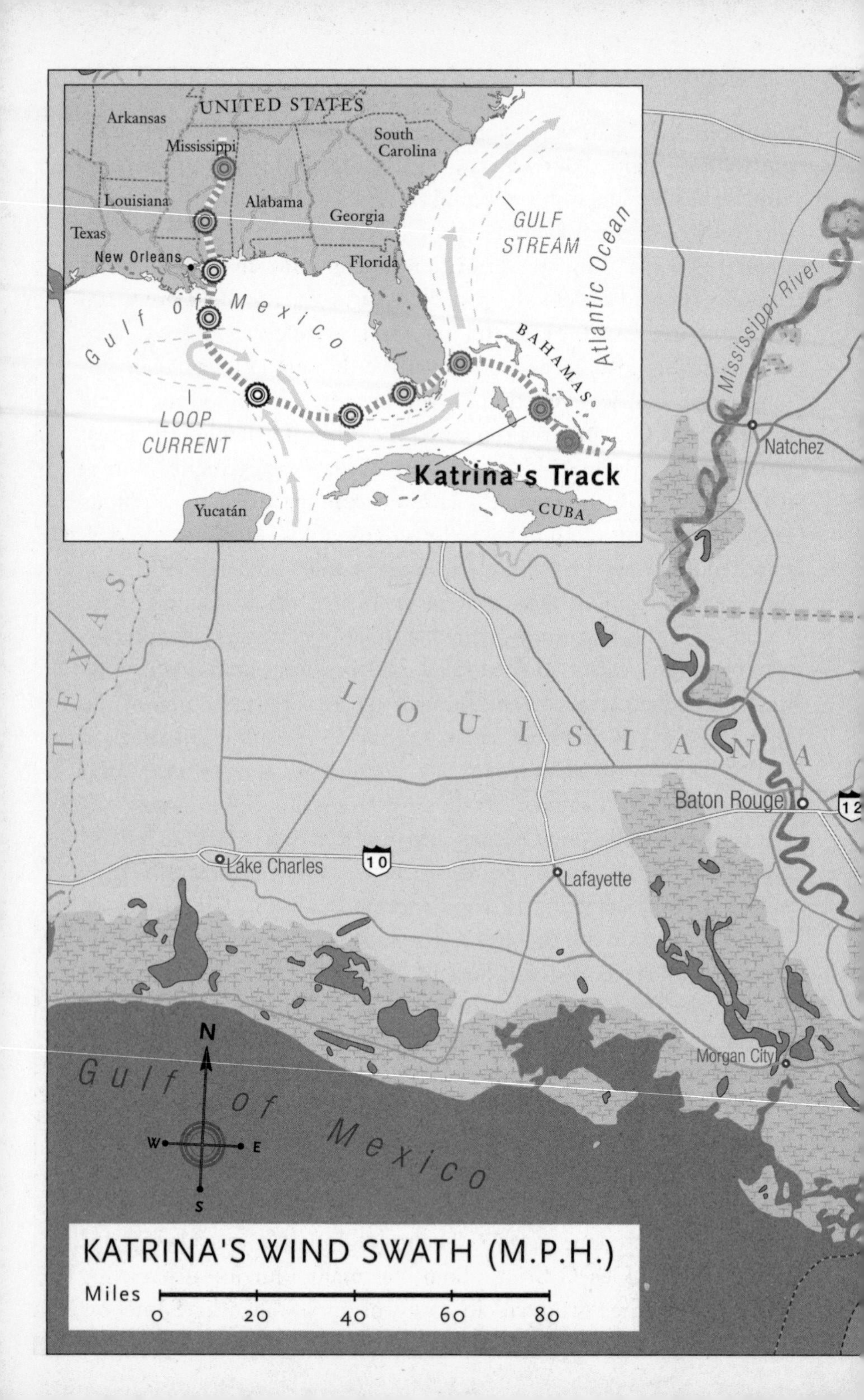

UNITED STATES
Arkansas
Mississippi
South Carolina
Louisiana
Alabama
Georgia
Texas
New Orleans
Florida
GULF STREAM
Atlantic Ocean
Gulf of Mexico
BAHAMAS
LOOP CURRENT
Katrina's Track
Yucatán
CUBA
Mississippi River
Natchez
TEXAS
LOUISIANA
Baton Rouge
Lake Charles
10
Lafayette
Morgan City
N
W
E
S
Gulf of Mexico
KATRINA'S WIND SWATH (M.P.H.)
Miles
0
20
40
60
80

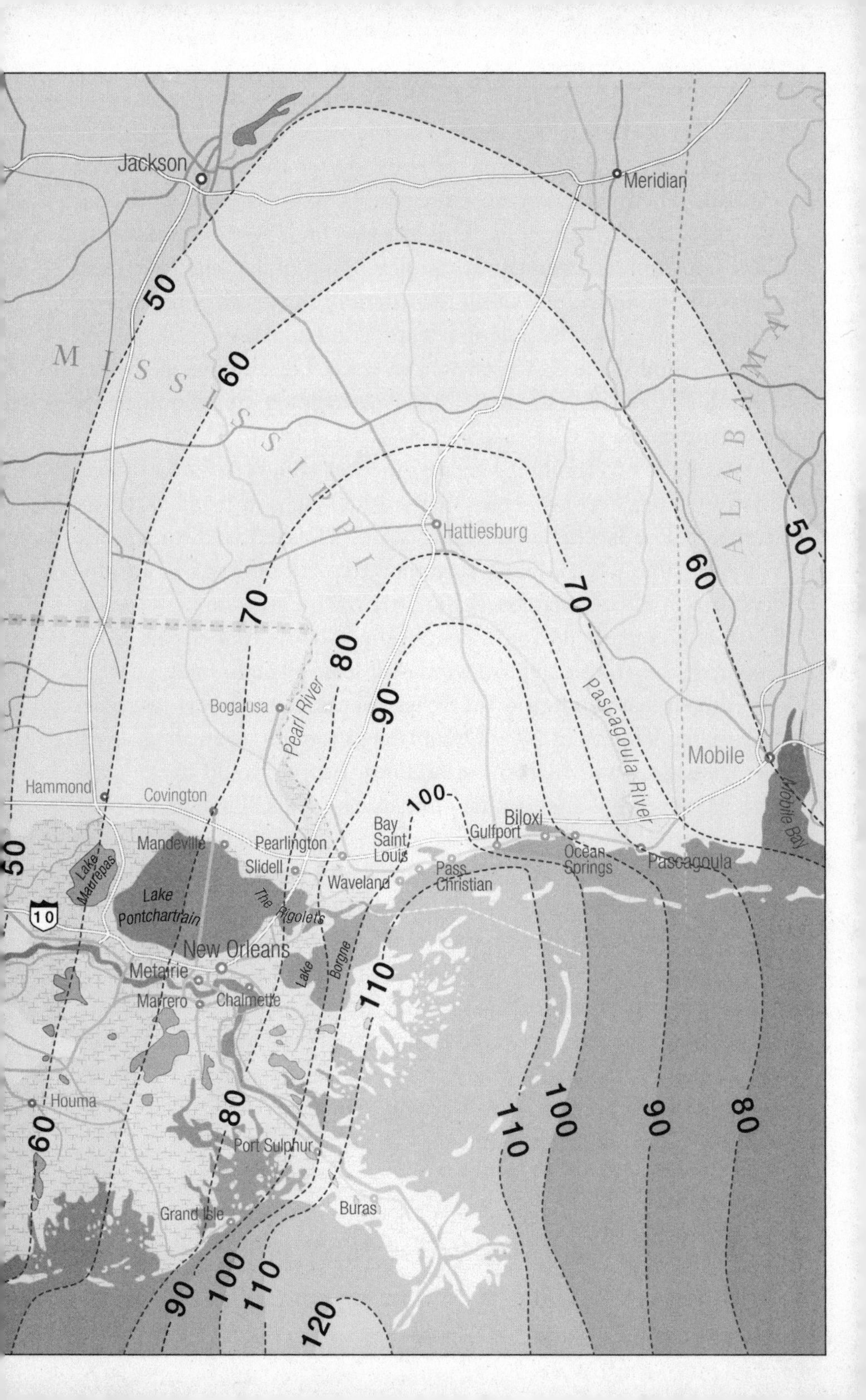

Jackson
Meridian
MISSISSIPPI
ALABAMA
Hattiesburg
Bogalusa
Pearl River
Pascagoula River
Mobile
Mobile Bay
Hammond
Covington
Mandeville
Pearlington
Bay Saint Louis
Gulfport
Biloxi
Ocean Springs
Pascagoula
Slidell
Waveland
Pass Christian
Lake Maurepas
Lake Pontchartrain
The Rigolets
10
New Orleans
Metairie
Marrero
Chalmette
Lake Borgne
Houma
Port Sulphur
Grand Isle
Buras
50
60
70
80
90
100
110
120

CNN had worked out a deal with the chancellor of LSU, Sean O'Keefe, who asked all of us to be handy for that network, which was fine. The request was accompanied by a note reminding us not to make landfall predictions. This was also fine. We never did that.) That four-minute interview at the beginning of day one gave me a great opportunity to talk about the extent of the flood, potential size of the rescue effort, the potential water contamination, the expected damage—and the need to restore the coast. Get the science before the public. Given the screwups that were to come, this proved especially important.

Driving to my daughter Vanessa's mother's house in Baton Rouge, I had to climb over two trees that crashed right in front of me on Highland Road. This was getting scary. I flashed back to a scene from my childhood, when a jacaranda tree fell on a car in a thunderstorm in Pietermaritzburg, the little town I grew up in. That car was totally squashed. Some people died. I was four at the time. Downed trees translate into a wind of at least 40 mph. Imagine what was already happening on the coast, where Katrina was just then making landfall, or in New Orleans eighty miles away, or over on the Mississippi and Alabama shorelines. People would die, I knew that. Meanwhile, Vanessa was sound asleep. Kids. The wind picked up dramatically as I drove east—alone (I was about the only car around)—with the idea of getting some quick sleep at home, but this proved nearly impossible. The wind was pretty loud, and I finally got up to take a look. It's amazing to see a 160-foot pine bend 40 feet or 50 feet, but our house is protected by quite a few acres of big trees—a healthy windbreak, so I thought we'd be fine. Then I heard a loud crack, and a cherry tree crashed onto a fence. Soon, a large oak missed the house by 6 feet, a magnificent magnolia by 8 feet. Again I needed the four-wheel drive to climb out the driveway and head back to work. It was time.

FOUR

LEVEES LITE

Everyone has seen the dramatic pictures: freighters on the Mississippi River towering above tourists enjoying their chicory coffee and beignets at Café du Monde in the French Quarter and, closer to Lake Pontchartrain, expensive homes hunkered in the shadow of what looks like an elevated canal of some sort across the street, protected by a very ordinary-looking concrete wall. The whole world must know now that 95 percent of New Orleans proper is below sea level, at an average depth of five feet. Most people don't know that the famous "bowl" is actually five independent bowls in the metropolitan area, each encased by its own set of levees.

One bowl includes the easternmost areas of the city and the outlying communities of Orleans Parish between Lake Pontchartrain and the Intracoastal Waterway: the Orleans East bowl, the deepest of them all. A second includes the easternmost areas of the city directly south of the first bowl, tucked between the Mississippi River Gulf Outlet waterway and the Mississippi River, and including the Lower Ninth Ward in Orleans Parish and a part of St. Bernard Parish that also harbors a large swatch of damaged marshlands: the St. Bernard Bowl. The western boundary of both of these bowls is the Industrial Canal that connects the lake and the river.

The third bowl extends from that large canal west to the 17th Street Canal, and from the lake to the river: the Orleans Metro Bowl, in my vernacular, the heart of the city. The fourth one reaches from the 17th Street Canal to the airport farther west, also from the lake to the river: the East Jefferson Bowl. (South and west of the 17th Street Canal the separation between these two bowls is the

Lake Pontchartrain
Pontchartrain Causeway
N
W
E
S
17th Street Canal
Orleans Ave. Canal
London Ave. Canal
Gentilly
Lakeview
Metairie
City Park
10
610
Mid-City
Memorial Medical Center I
CANAL ST.
French Quarter
ST. CLAUDE AVE.
S. CLAIBORNE AVE.
CARROLLTON AVE.
Broadmoor
Charity Hospital
Superdome
Central Business District
Algiers
Convention Center
Crescent City Connection
Tulane University
Memorial Medical Center II
MELPOMENE ST.
Audubon Park
Garden District
ST. CHARLES AVE.
90
WEST BANK
JEFFERSON
ORLEANS

THE FLOODING OF NEW ORLEANS
Break or breach in levee
FLOODING DEPTHS
18 – 16 feet
15 – 11 feet
10 – 6 feet
5 – 1 feet
No flooding
Miles
0
2
4
Industrial Canal
New Orleans East
10
Mississippi River Gulf Outlet Canal
Intracoastal Waterway
Lower 9th Ward
Jackson Barracks
Arabi
Chalmeette Battlefield
Chalmette
Mississippi River
LEVEES TO THE EAST ALMOST COMPLETELY COLLAPSED

Metairie Ridge, which is actually the levees of an old distributary of the Mississippi). The fifth bowl is on the other side of the Mississippi River, the south side but called the West Bank because it *is* the west bank farther upriver. These outlying communities in Orleans and Jefferson parishes are somewhat protected by another set of levees ("somewhat" because the system is not completed, and won't be for years—and should not be, as I'll explain in the final chapter).

Those are the five main bowls. In addition, a long, narrow wedge of communities in Plaquemines Parish about forty miles southeast of New Orleans is protected by levees on the east side of the Mississippi. Downriver from there is another protected wedge on the west side of the river. Thirty miles directly south of New Orleans, another large area of dense development around Bayou Lafourche is encircled by levees. The grand total is about 500 miles of levee protection in southeast Louisiana (350 of them in Greater New Orleans). Prior to Katrina, the Army Corps of Engineers was working

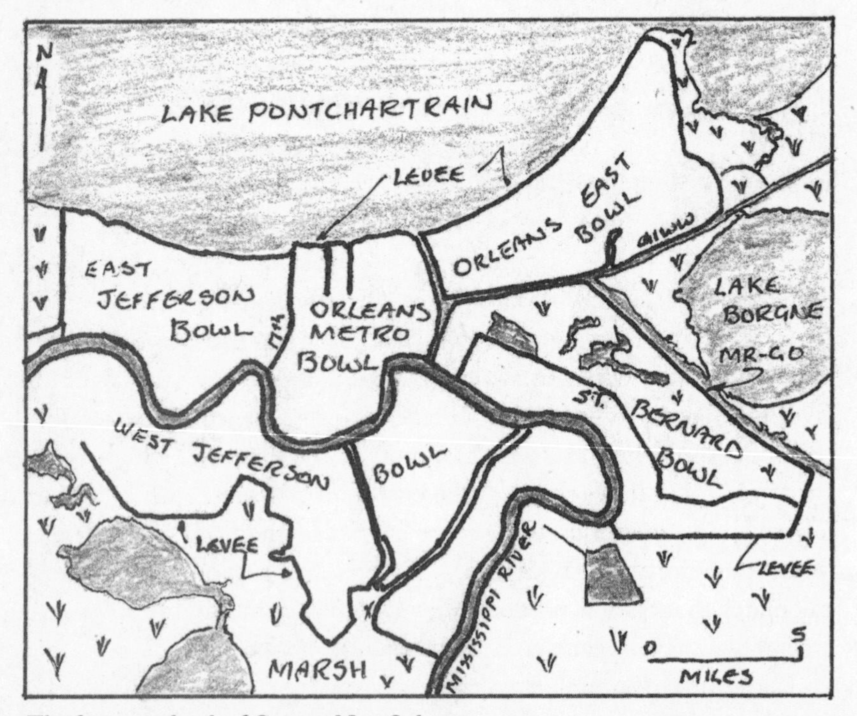

The five main bowls of Greater New Orleans.

on four major projects. One was the levees around Lake Pontchartrain; another was the Morganza-to-the-Gulf levee, designed to protect Morganza, Houma, and four other towns at a cost of well over $1 billion. Still in the design stage, this structure would be seventy-two miles long, nine to fifteen feet high, and state-of-the-art, a complex "leaky levee" whose floodgates and culverts would allow water to flow back and forth and replenish the marshes. That was the idea, and three years ago Terrebonne Parish levied, with popular approval, a quarter-cent sales tax to handle its share of the cost. Those taxpayers now have something over $10 million set aside. Not quite enough. In any event, all such plans are now up in the air, obviously.

Rainfall aside, the specific sources of flooding water that necessitate all of these levees are the Mississippi River, Lake Pontchartrain, and storm surge rolling in from the Gulf of Mexico to the east and south. Today, the operative problem is the last two—the lake and the Gulf—because the mighty river is now effectively held in check. It has not seriously flooded New Orleans or anywhere else in Louisiana since 1927, and it is not likely to again. Upstream from the city four different spillway systems, beginning with the Old River Control structure that diverts water into the Atchafalaya River one hundred miles to the northwest and ending with the Bonnet Carré Spillway just west of the city, can, in an emergency, dump fully half of the river's flow, as much as 1 million cubic feet per second. The levees along the river are 300 feet wide at the base and 50 feet at the top. They aren't going anywhere. They are built to an elevation of 25 feet above sea level, while the average annual high-water mark for the river is 14 feet. An unbelievable amount of water would be required to fill up those last 11 feet of volume in the channel, which is 600 yards wide as it flows through the city at an average depth of 90 feet. There is a tremendous margin for error designed into these levees. The Army Corps of Engineers, which is solely responsible for them, calculates that they'll protect New Orleans from anything less than an eight-hundred-year flood. That could be about right.

Local provocateurs do enjoy speculating that a tipsy captain could miss the hairpin right-hand downstream turn at the French

Quarter and smash his ship through the levee, but it would take a hell of a bash. On a Saturday afternoon in 1996—December 14, 2:30 P.M.—the *Bright Field*, a 763-foot freighter registered in Liberia (where many are, for legal purposes) and loaded with 56,000 tons of corn bound for Kashima, Japan, via the Panama Canal, rammed into the dock with the RiverWalk mall and hotel complex, not far upstream from the hairpin turn. No deaths, but 116 people were injured, with millions of dollars of property damage. At that location, a dock protected the levee, but the crash did make people wonder about the worst case. Really, though, the scarier scene for engineers is the collection of pressure-release wells on the west side of the Mississippi not far north of Baton Rouge. When the river is low, nothing is happening, but when it's up, these wells are bubbling with river water seeping under the levee, which right here happens to sit on a sandy base, which is highly permeable. When the river is high the pressure will push water through this sand and out the other side. We call these little eruptions sand boils, and without these wells to relieve that pressure, the levee here could fail, with a serious flood fight to follow. There's another collection of sand boils on the east side of the river not far from the LSU campus. Pumps move this unwanted water back into the river. Elsewhere, the Corps has laid down concrete mats, in an attempt to control erosion, but some deep holes have developed nevertheless.

The river levees do bear watching, and they are watched, but, all in all, the river is not the problem. The danger to New Orleans and vicinity, proved time and again and most catastrophically by Hurricane Katrina, has been the much less substantial hurricane levees along Lake Pontchartrain, the Industrial Canal, the Intracoastal Waterway, the Mississippi River–Gulf Outlet, the Harvey Canal in the West Bank, and elsewhere. None of these levees—none—are (or were) anywhere near as high or intimidating as those along the Mississippi River. The levees along Lake Pontchartrain range in height from 13 feet to 18.5 feet. (The difference principally reflects the designers' fear of wave "run-up"; that is, lots of waves pounding the shoreline and thereby significantly raising the effective level of the surge.) The other levees around the region range in height from a mere 5 feet to 17 feet. Extensive history and our storm-surge models produced long before Katrina show insufficiencies in many different areas, each a disaster waiting to happen, even without actual failures such as occurred during Katrina. With every levee holding intact and doing its job as built, parts or essentially all of New Orleans will still go under when a slow-moving major hurricane following any one of numerous scenarios hits the region. The details of the surge flooding disaster would depend only on the strength, forward speed, and direction of approach of the storm. With Katrina, the ADCIRC surge models pinpointed the primary problem as the levees along the Intracoastal Waterway and the Industrial Canal, which would be overtopped by the storm surge rolling in from the Breton and Chandeleur sounds and pushed up the Funnel, a feature shaped by the waterway and another shipping channel, the Mississippi River–Gulf Outlet (MR-GO).

MR-GO, as it is always labeled (sarcastically pronounced "Mister Go"), was built by the Corps to provide shipping with a straight shot into the Gulf of Mexico, a 76-mile route that cuts 40 miles off the trip down the winding channel of the Mississippi River. This route was originally authorized to be 650 feet wide at the surface, 500 feet at the bottom, 36 feet deep. It required removing more dirt than did the Panama Canal—hard to believe—and from the earliest planning stages in the late 1950s it was challenged by a host of opponents—including the Corps, because it did not remotely satisfy any of its cost-benefit analyses. Keep trying, Congress replied. The Department of the Interior stated that "excavation could result in major ecological change with widespread and severe ecological consequences." That's exactly what has happened. As a result of erosion, the channel in some stretches is now three or four times as wide as the design specification. Contiguous marshlands have been severely damaged, if not ruined. The canal feeds saltwater directly from the Gulf of Mexico into freshwater marshes and swamps and has effectively killed thousands of acres of wetlands, which are now just open water marked by the trunks of the odd dead cypress trees.

As I'll discuss in detail later, levees protected by healthy marshes are much less likely to fail. Simply put, MR-GO has devastated the

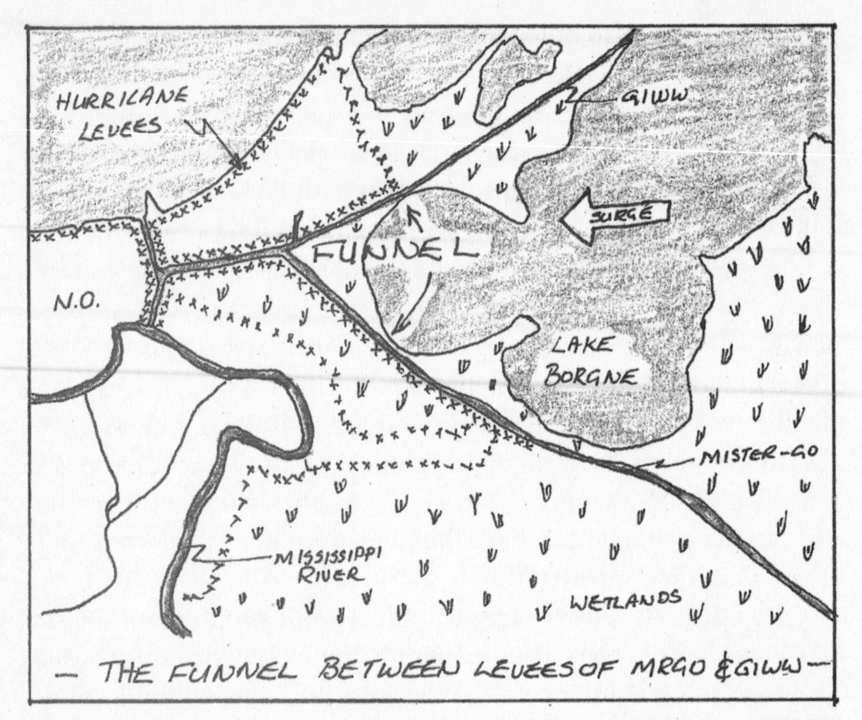

The Funnel between the levees of MR-GO and the Intracoastal Waterway [GIWW].

protective marsh structures immediately east of New Orleans, and by ruining these adjacent marshes it has made its own levees that much more susceptible to erosion and failure. In 1998, the St. Bernard Parish Council unanimously called for closing the channel. The following year Sherwood "Woody" Gagliano, an independent consulting coastal expert, presented a paper at a meeting organized by the University of New Orleans to consider the MR-GO question. Woody concluded, "The Mississippi River–Gulf Outlet, since its construction in 1965 as an alternative route for oceangoing vessels into the Port of New Orleans, has caused increased storm-surge vulnerability to developed areas of St. Bernard and Orleans parishes and extensive environmental damage to a vast region. Greatest impacts occur in St. Bernard, Orleans, and Plaquemines parishes, in that order. The channel is a serious threat to public safety and an environmental threat to the region."

No wonder it's also called Storm Surge Alley, or Hurricane Alley. Shut it down! Rebuild those marshes! Almost everyone agrees, but it still hasn't happened, even though only five ships a day, maximum, use the thing. Following Hurricane Ivan in 2004 the Corps even spent $17 million dredging it for the sole benefit of those lonely ships. Generally, annual maintenance costs vary from $13 million to $37 million.

The impact on the marshes is one problem with MR-GO. The second, more immediate problem, is its levees, because storm surge pushing across shallow Lake Borgne from the east is constrained by these MR-GO levees to the south and, to the north, by the long-standing levees of the Intracoastal Waterway. Initially ten or more miles apart, these two channels meet, and when they do the water building between their levees is squeezed into a single channel—the Funnel—only 260 yards wide, constrained by levees 14 feet to 16 feet high. Surge warrior Hassan Mashriqui has studied this phenomenon with zealous attention. His series of surge hydrographs and other velocity plots demonstrate without a doubt just how bad the "funnel effect" is right here. In concert with the denuded marshes, it could increase the local storm surge hitting the Intracoastal Waterway by 20 percent to 40 percent—a "critical and fundamental flaw" in the system, in Mashriqui's phrase. I remember the scene when he and Paul Kemp asked me to look at this data. They knew it was critically important, and so did I, because the levee designs and heights out here did *not* take this funnel effect into account. They weren't high or strong enough.

We have to get this news out, I said. Mashriqui and Paul started with a poster at a conference, and then we sought every other possible venue to warn people about this inherent weakness in the levee system. In January 2005, Walter Brooks, executive director of the Regional Planning Commission of the New Orleans Metropolitan Area, asked me to give a talk about hurricane surges and vulnerabilities at their offices in New Orleans. The surge amplification caused by the Funnel was a featured point, and during the ensuing discussions I was asked if we at LSU could come up with a conceptual plan to build a structure to protect the area. I said yes, of course. Jefferson Parish president Aaron Broussard said they couldn't wait for the

Corps, that the parishes would have to fund this effort themselves. The two Corps officials who were there said the Corps could also develop plans, but I got the impression that these parish presidents meant business. Walter Brooks asked me to return and give a longer talk that would be taped and aired on local cable TV. I was going to be overseas, so I suggested that Mashriqui give the talk. The Funnel was really his baby, anyway. He did a great job, and right before Katrina struck we were getting ready to draw up a proposal for the conceptual planning study.

The Funnel is six miles long. To the west, it—and the water in it—"T" into the Industrial Canal. It's hard to believe, if we step back to think about it. The federal powers that be had inadvertently designed an excellent *storm-surge delivery system*—nothing less—to bring this mass of water with a simply tremendous load—potential energy—right into the middle of New Orleans. If during any given storm the levees along the Intracoastal Waterway and MR-GO have not already been overtopped—or even if they have been, in a big storm like Katrina—something has to give at this critical intersection, and if the levees are anything less than optimal, it's not going to be the water.

That's what happened on the Monday morning when Katrina struck. At 6:10 A.M. the hurricane made landfall at the small town of Buras in the farthest reach of the Mississippi Delta. Katrina had weakened considerably overnight. By dawn, she was a minimal Cat 3 storm, with highest winds of 112 mph. Buras and vicinity were nevertheless devastated, of course. Forty miles to the north, where the winds were still a mere 60 mph, the first flooding of residential areas in the greater New Orleans area had already begun. The storm surge building on Lake Borgne east of New Orleans would peak at 18 feet at about 8:30 A.M., but four hours earlier the huge waves on top of the surge, driven by the winds from the east, made fairly quick work of certain stretches of the MR-GO levees, which were overwhelmed and in some cases destroyed. Water poured into the lower areas of the bowl between the Intracoastal Waterway and the Mississippi River, including Chalmette, Meraux, and Violet, with the Lower Ninth Ward farther to the west in serious jeopardy.

At about 6:30 A.M., the surge of fourteen feet to seventeen feet in the Funnel proper—the confluence of MR-GO and the Intracoastal Waterway—overtopped the levees on both sides. To the south this water poured into the communities that were already taking on water through the MR-GO breaches, now including the Lower Ninth Ward. To the north it poured into the neighborhoods in the adjacent bowl in Orleans Parish, between the Intracoastal Canal and Lake Pontchartrain. (Some of these neighborhoods had already been taking in some water from the earliest of all the breaches—between 4:30 A.M. and 5:00 A.M.—at the junction of the CSX Railroad and the northern arm of the Industrial Canal, right next to Interstate 10. The metal gates where the railroad tracks pass through the I wall of the levee were not working, apparently because of a prior derailment, and the pathetic sandbags in their stead gave way early on both sides of the canal. We know this because a nearby gauge measuring the water level recorded a rapid drop from nine feet to four to five feet above sea level. Water poured into the Orleans East bowl and the Gentilly section of the Orleans Metro Bowl to the west. While this breach was not a huge flood maker, it would have frightened local residents and, we can hope, alerted them to head for high ground—that is, the rooftops.)

At about 6:50 A.M., the surge coming through the Funnel hit the T at the Industrial Canal; some of this water was forced to the right, or north, and poured into Lake Pontchartrain, which was then six feet lower than the surge. The rest was forced to the left, or south, where it was blocked by the closed locks that connect the waterway and the Mississippi River. (The locks were closed in order to separate the surge in the river from elsewhere.) As Mashriqui's model run had predicted, the levees along both sides of the Industrial Canal, from the river to the lake (a distance of five miles), were now overtopped. Water poured into the two bowls to the east, which were already taking on water, and now also into the bowl to the west—the Orleans Metro Bowl.

Alas, this overtopping did not relieve enough of the pressure on the flood walls. At 7:45 A.M., give or take not many minutes, two different sections of the levee along the eastern side of the southern end of the Industrial Canal—a total of about four hundred yards—

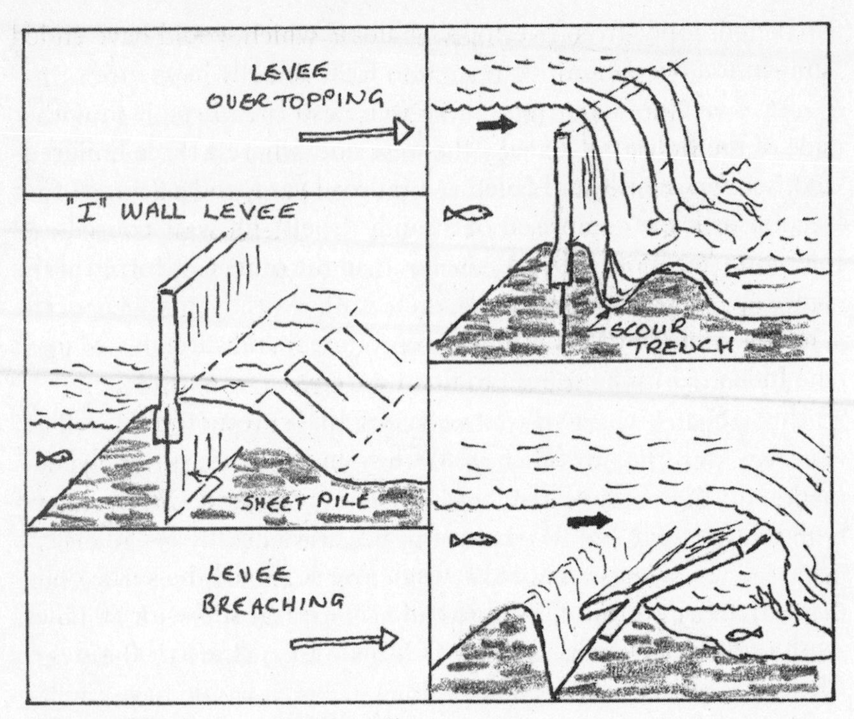

A breached levee versus an overtopped one.

abruptly collapsed. A wall of water, literally, exploded into the Lower Ninth Ward, with truly catastrophic results. Because these breaches occurred *before* the peak of the surge at 8:30 A.M., this neighborhood four feet *below* sea level was drowned by water as high as fourteen feet *above* sea level. That eighteen-foot head of water explains the total devastation of the homes within the three blocks of the breaches—devastation very like that on the Mississippi shoreline at just about the same time. Hundreds drowned in the space of about ninety minutes. Thousands were scrambling for their lives.

With overtopping, water is simply too high for the levee and flows over it. A breach is a rupture, a failure of some sort. A levee can be overtopped but never breached, or it can be breached even though it is never overtopped. The *breaches* along the Industrial Canal completely mooted our storm surge models for the St. Bernard Bowl, which of necessity assumed no such breaches and therefore predicted floods from overtopping alone, which would have ended shortly and been minor compared to what actually happened.

At about the same time, there were also breaches on the other side of the Industrial Canal—the west side, where a three hundred-foot section of an I wall failed at a railroad yard, and a seventy-foot section of levee comprised of a sandy "shell" fill was scoured and blew out. Yet another break, smaller than the others, occurred at the junction of a soil berm and a concrete wall. Thus the integrity of the Orleans Metro Bowl was now compromised, and surge water from the Industrial Canal started to flood the city.

Just about the time the levees along the Industrial Canal failed, the eye of Katrina was passing to the east on its way to its second landfall on the Mississippi coast, now only as a Cat 2 storm with sustained winds of 98 mph. So the winds were way down, but the storm surge was not. Twenty-five feet to 28 feet above sea level, the surge obliterated almost everything within half a mile of the shore for 50 miles, with serious damage as far east as Mobile Bay, almost 100 miles to the east of the second landfall.

Extremely important question: Since Katrina was barely a Cat 3 storm at landfall on the Louisiana coastline and only a Cat 2 when the eye crossed the Mississippi coastline, why the record high surge? The answer is threefold. First, the momentum of the surge was established while Katrina was definitely a Cat 4 and a Cat 5 out in the Gulf of Mexico. As LSU's Elizabeth English put it, "There was essentially a lot more momentum in the water than there was in the wind." Second, the surge could maintain this momentum because the wetlands in its path are tattered, as are the Mississippi barrier islands. A third factor may have been the levee along the east bank of the Mississippi River. At landfall just after 6:00 A.M., a twenty-one-foot surge was building against this barrier from the east. Since the levee height is eighteen feet above sea level, it was soon overtopped. The surge immediately filled the river itself, in effect, then overtopped the levee on the west side and flooded the surprised residents who stayed behind over there. This surge was so powerful that even as it overtopped the Mississippi levees, therefore losing some of its head and volume, it *continued to build* to the north. Just an incredible

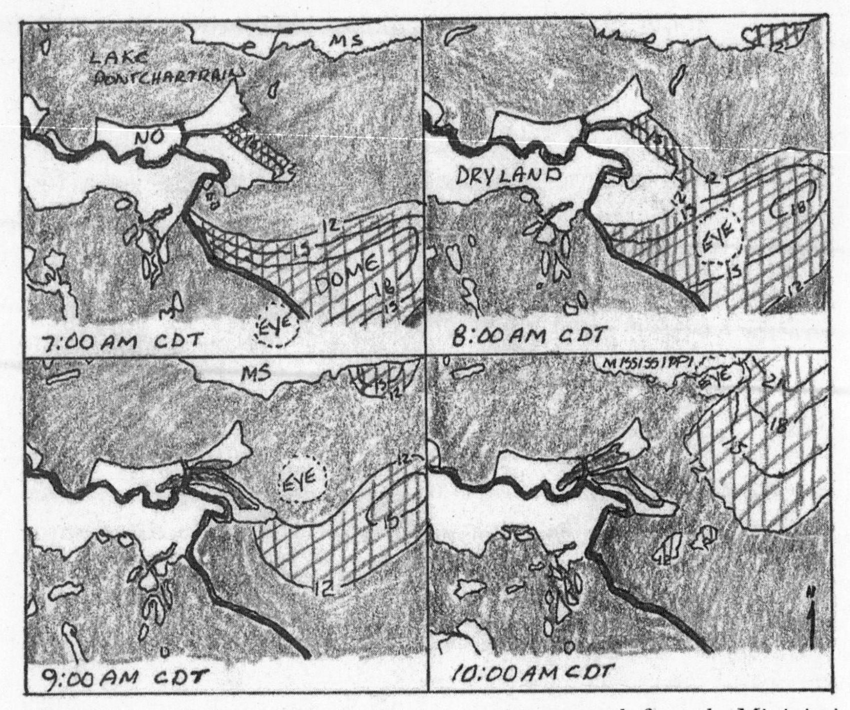

These four panels depict the surge dome as it migrates north from the Mississippi River levee to the Mississippi coast. Note how the surge dome forms ahead of the eye of the storm, is picked up by the eye, and then moves to the Mississippi coast.

surge, but hurricane guru Mark Schleifstein of the *Times-Picayune* has proposed a brilliant hypothesis by way of further explanation. Mark suggests that the levee on the east side of the river is so massive a barrier that it acted as a dam for that quadrant of the surge. As the eye of Katrina moved north, so did this dome of water. By 8:00 A.M., the eye had migrated over this dome, and the westerly winds on the backside of the eye then shoved the dome toward the east, away from the levee. The dome was then caught by the counterclockwise rotation, and at 10:00 A.M., as the eye made landfall on the Mississippi coast, so did the storm surge, augmented by this extra dome of water created by the levee along the Mississippi River. I wish I could say that this fascinating idea was mine, because it rings true to me as a partial explanation of the supersurge in Mississippi. We'll have to investigate further, of course. If Mark's theory proves correct, we have this terrible irony: The levees that protected adjacent areas from the Mississippi River exacerbated the storm surge on the Mississippi coastline to the north.

By 10:00 A.M. the eye of Katrina was well to the northeast of the New Orleans area, and the storm surge in the Funnel and the Industrial Canal was dropping as the now westerly winds pushed the water back to the east. Water quickly started draining out of the Lower Ninth. (It stabilized three days later at plus-five feet above mean sea level, then fell slowly to about plus three on Saturday.) In the bowl to the north, Orleans East, between the Intracoastal Waterway and Lake Pontchartrain, the floodwaters were also stabilizing until yet another pulse of water exacerbated this flooding, thanks to the fact that the concrete flood wall that extends for about one mile behind the Lakefront airport on Lake Pontchartrain is two feet lower than the earthen levee on either side. This aberration in levee height is quite bizarre. The earthen levee was not eroded or overtopped, even though the lake level was one foot from its crown, but the lower section with the concrete flood wall was overtopped for about three hours. What team of engineers designed the concrete levee two feet lower than the adjacent earthen counterpart? Did these engineers believe that water won't flow over a concrete wall?

The storm surge from Katrina that pushed the mass of water into the two easternmost bowls of New Orleans also pushed an exponentially larger volume of water to the north across Lake Borgne, across skimpy marshes, and through two tidal passes into Lake Pontchartrain. Here the water was trapped, and our surge warriors have modeled this basic scenario for about fifty storms. With these hypothetical storms, as the wind continues to push in the surge from the east, the lake rises, floods the low sections of Slidell on the north shore (which, for the most part, are unprotected by levees), and pounds the lakefront levees of New Orleans, twenty-five miles to the southwest. As the storm proceeds to the north, passing the lake on the east, the wind shifts to a more northerly direction, shoving even more water toward the southern rim of the lake and really challenging those levees, which rise to a height of fifteen feet to eighteen feet above sea level. At this moment, the life of New Orleans depends entirely on those barriers. What happens next depends on the details of the storm—mainly, the *exact* course and speed of the storm prior to reaching the lake, and the strength and direction of the winds on the lake. Several modeled scenarios produce a surge that simply overtops all of the levees protecting New Orleans and drowns the city completely—no failures, no collapses, just wholesale overtopping of levees that are not high enough.

That was *not* the scenario for Katrina. Mashriqui's last model runs on Sunday predicted that the lakefront levees would suffice. Those models were working with the prediction that Katrina was a fast-moving Cat 4 storm that would lose a lot of steam after landfall and track to the east of the city, leaving Lake Pontchartrain exposed only to the left-hand side of the advancing storm, where the wind speeds would only be that of a Cat 1 storm. That's what happened, and those levees did suffice, with the exception of the section directly inland from the Lakefront Airport, as I just noted.

What did not suffice were three sections of the levees along two long drainage canals that extend from Lake Pontchartrain deep into New Orleans. These walls themselves were never overtopped by the storm surge. They were high enough.

Were the walls as high as officially stated? Over the following months, some observers would suggest that they weren't, thanks to subsidence. The role of subsidence is definitely an important one for the Katrina story, as we'll see, and accurate survey data on the heights of the various levee systems in the greater New Orleans area is scarce. Based on the best data we have, however, levee subsidence had very little if any impact on the Katrina disaster. There is some evidence of differential subsidence with some levees, where the soil under a given section has compacted and subsided faster than adjacent areas, but none of these sections that we know about were overtopped. So subsidence is part of the levee landscape, but it does not appear to have exacerbated the New Orleans flooding. Instead, three sections of the levees on two of the drainage canals—a total of about five hundred fatal yards—just collapsed, catastrophically, drowning much of the Orleans Metro Bowl.

Drainage canals are required all over New Orleans for the simple reason that it rains here—a lot—and this water would collect in the bowls if it weren't pumped up and out into Lake Pontchartrain by way of these canals. On the other hand, some of these canals, with the pump stations located deep within the city, could thereby also serve in a storm to bring the lake into the city, so they must be equipped with their own hurricane levees. For the most part, these walls are protected from the violent wave action on the lake that batters the lakefront levees—a key advantage, though not key enough, as it turned out.

Enter Louisiana politics. The levees along the drainage canals (and throughout the region) are built by the Army Corps of Engineers, with the construction costs shared by the local levee districts, which align with the parishes. These districts are then responsible for maintenance and any other improvements they desire and are willing to pay for. Greater New Orleans is one metropolis but two parishes and two levee districts: Orleans Parish in the east, Jefferson Parish in the west. The north-south dividing line is Seventeenth Street, just as the 17th Street Canal is the dividing line between the two westernmost bowls. This is a strange setup, and it can't possibly be ideal. Complicating matters further, the levee districts are responsible only for the levees themselves. Parallel water and sewerage boards are responsible for the canals per se, and the pumping stations. Assessing this scheme and the number of boards and authorities involved, one might suspect that political patronage has played a role over the years, and might by now be deeply rooted.

In Jefferson Parish to the west the four state-of-the-art pumping stations are at the mouth of each canal, right by the lake, where they function as part of the defense against storm surge, with well-engineered and substantial concrete levee walls connecting the pump stations to the earthen levees along the lakeshore. The water in these drainage canals is not an unimpeded extension of the water from the lake. In Orleans Parish, however, we have a very different and more vulnerable system, on its face. The pump stations for the London, Orleans, and 17th Street canals are at the foot of each

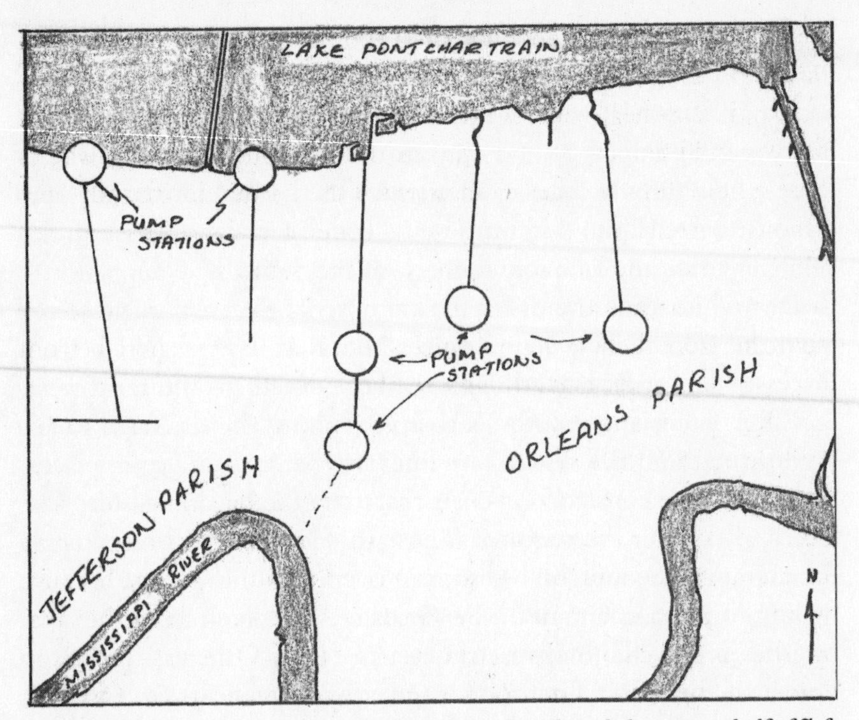

Locations of the major pump station for Orleans Parish and the eastern half of Jefferson Parish. Note the Jefferson pump station at shoreline, Orleans within the city, necessitating long parallel levees from pump station to lake.

canal, two or three miles inland. The water in these canals *is* simply an extension of the lake. The canal levees are therefore much more vulnerable to storm surge. A Dutch member of the postmortem team brought in by the American Society of Civil Engineers compared the two systems and asked about the one in Orleans Parish, "Why in the world would you invite the enemy deep inside your own camp?" Indeed.

Why the difference between the two parishes? I'll dig into this complicated issue further when I explain the investigation of the levee failures, but the difference partially reflects the simple fact that the Orleans pump stations were in the right place early in the twentieth century, when they were originally built at the edge of town. As the city expanded north toward the lake (which was possible only as pumping drained the swamps), the stations ended up deeper and deeper inside the burgeoning metropolis. In suburban Jefferson Parish, on the contrary, where residential and commercial development has been relatively recent, it was obvious that these neighborhoods would extend all the way to the lake, and that the lake is where the pumps belonged. In fact, it was the building of the lake levees and initial pump stations that allowed Jefferson Parish to develop and expand so rapidly in the 1960s and 1970s.

During Katrina, the difference became tragic. As the eye of the hurricane approached the entrances into Lake Pontchartrain from the east, the winds along the south shore swung to the northwest, pushing the violently turbulent surge, topped by ten-foot waves, against the lakeside levees. The surge itself was eleven feet in places, exactly what our ADCIRC storm-surge models had called for. The levees held. There was some wave splash over a few sections, but no overtopping. In Jefferson Parish, that is the end of the story, because these four main inland drainage canals are protected from the lake. In Orleans Parish, however—the heart of New Orleans—the surge pushed into the three main drainage canals—17th Street, Orleans, and London Avenue—impeded only by a bridge across each canal close to the lake. I say "impeded," but in fact the bridges had no negative effect on the surge heights at all, serving only to dampen waves that may have tried to propagate up the canal from the lake, as well as to slow down any water flowing in or out of the canal.

The storm surge in the lake peaked at 9:00 A.M. Monday. Less than half an hour later, when the water would have been at least four feet below the tops of the levees, a one-hundred-yard-wide section of the levee on the eastern side of the London Avenue Canal near the Mirabeau Bridge ruptured—failed catastrophically. How catastrophically? The home directly in front of the center of that breach was shoved ninety feet across the street. I managed to get some excellent home-video footage taken just after the breach occurred. It shows the water pouring in, over a mile and a half from Lake Pontchartrain and deep in the heart of the city. These waters coursing through this breach soon met the waters from the breaches and overtopping along the Industrial Canal levee a little over two miles to the east. The waters also served to lower the level in the canal, and the storm surge itself was also abating. So the head of pressure in that canal was now significantly less than it had been just half an hour earlier. Nevertheless, at about 10:30 A.M., when the surge was only seven feet above sea level, a second section of these London Avenue levees failed, this one on the west side and just south of the Robert E. Lee Bridge, less than a mile from the lake. An eight-to-ten-foot head of water poured into the neighborhoods of Lakeview to the west.

Not quite two miles to the west the levees of the Orleans Canal held, no doubt helped by the fact that the flood embankment fronting the pump station at the foot of the canal was overtopped for at least two hours. This embankment, about four hundred feet long, is six feet lower than the flood walls along the canal. The flood walls here are missing! This overtopping caused significant and early flooding at City Park, but it also lowered the pressure on the flood walls between the pumping station and the lake. In effect, Orleans Canal was a spillway for two hours, and the flood walls held. But another mile farther west, at the 17th Street Canal, one hundred yards inland from a new hurricane-proof concrete bridge at the Metairie Hammond Highway, the levee on the east side failed at approximately 9:45 A.M., a little earlier than the second breach on the London Avenue Canal. This breach flooded Lakeview, which was now taking on water from two sides.

Water would pour through the three breaches of the drainage canals in the New Orleans Metro Bowl for more than sixty hours, until early Thursday morning, when the water in this bowl equalized with the water in the lake at about three feet above sea level, and with the average home in the flooded sections standing in six to nine feet of water.

It all sounds so clinical, and it *was* clinical after the fact, as we set about laboriously putting together the narrative for all of the breaches around the city, collecting eyewitness reports, the rare time-stamped amateur video footage, professionally shot footage (almost all from helicopters), and even battery-powered and other clocks that were stopped by submersion in water. The idea was to correlate the reports and actual evidence with our hydrographic predictions, in order to establish the most complete possible forensics. What had caused these catastrophic levee breaches? This was the only way to find out. An important advantage of the ADCIRC program is that we can generate a flood hydrograph that depicts the water level at any given location over time. This graph generally has a bell-shaped curve—water stable, water rises, water recedes—with the slope and exact shape of the curve depending upon the location chosen. The hydrographs tell us when the surge peaked—vitally important to any good forensic study.

The following timetable summarizes the flooding of Greater New Orleans on Monday morning, August 29, with the arrival and passage of Hurricane Katrina:

4:30 A.M. or shortly before: The relatively minor breaches at the intersection of the CSX Railroad and the northern arm of the Industrial Canal send water east into Orleans East and west into Orleans Metro bowl. This flooding lasts about thirteen hours. Limited flooding into southern St. Bernard over the east bank of the Mississippi near Poydras.

5:00 A.M. give or take: The MR-GO levees in St. Bernard Parish begin to erode and fail, with water pouring into the St. Bernard Bowl, starting to fill the wetland basin that lies between the MR-GO and the state back levee that protects the homes. Flooding was continuous for days, until the levees were sealed and pumping started.

6:10 A.M. Katrina makes landfall at Buras, on the west bank of the Mississippi River. Storm surge overtops the levees on the east bank of the river, crosses the river, overtops the levees on the west bank, and sends additional water into these neighborhoods in Plaquemines Parish. Levees on both sides of the Gulf Intra Coastal Waterway are overtopped, some start to fail, initiating flooding of Lower Ninth and adding to flooding of Orleans East.

6:30 A.M. The levees in the infamous Funnel are overtopped, flooding the Orleans East Bowl and adding to the flood in the St. Bernard Bowl.

6:50 A.M. The levees on both sides of the Industrial Canal are overtopped, with water pouring into the Orleans East, St. Bernard, and Orleans Metro bowls.

7:30 A.M. Sections of the levees on the west side of the Industrial Canal at the railroad yard are breached. This flooding lasts for twelve to fifteen hours.

7:45 A.M. The levee on the east side of the southern end of the Industrial Canal catastrophically breaches in two sections, sending a wall of water into the Lower Ninth Ward and other neighborhoods of the St. Bernard Bowl. Serious flooding of St. Bernard occurring as the wetland bowl is now full and water is pouring over the back levee into the homes.

8:15 A.M. The embankment at the foot of the Orleans Canal is overtopped and floods City Park for about three hours.

8:30 A.M. The one-mile stretch of levee behind Lakefront Airport that is two feet lower than the adjacent earthen levee is overtopped by the surge from the lake. Floods for about three hours.

9:00 A.M. The storm surge from Lake Pontchartrain peaks in the three drainage canals in the Orleans Metro Bowl, with no overtopping of the levees.

9:30 A.M. Though never overtopped, one section of the levee on the east side of the London Avenue Canal near Mirabeau Bridge breaches catastrophically, flooding the Orleans Metro Bowl.

9:45 A.M. or thereabouts: Though never overtopped, one section of the levee on the east side of the 17th Street Canal breaches catastrophically, flooding the Orleans Metro Bowl.

10:30 A.M. Though never overtopped, one section of the levee on the west side of the London Avenue Canal breaches catastrophically, flooding the Orleans Metro Bowl.

It must be repeated: If the only sources of water in New Orleans had been the rainfall from Katrina (seven to ten inches), the predicted overtopping of the levees of the Intracoastal Waterway and the Industrial Canals, the overtopping of the Lakefront Airport levee, and the breach at the CSX railroad junction, the flooding in the Orleans East and St. Bernard bowls on the eastern side of Greater New Orleans would have been much less damaging, and the flooding in the Orleans Metro Bowl, the heart of the city to the west, would have been relatively insignificant. According to our latest calculations, 88 percent of the flooding in the Orleans Metro Bowl, by volume, was due to the breaches on the London Avenue and 17th Street canals. In Orleans East, 69 percent of the flood was due to breaches. In the St. Bernard Bowl, 92 percent. Thus, on average, *87 percent* of all the water that ended up flooding the greater New Orleans metro area was the result of levee failures that totaled less than 400 yards on the drainage canals and 650 yards on the Industrial Canal. It seems terribly unfair, and it is, but how large was the damaged area on the underside of the wing that doomed the space shuttle Columbia? One square foot.

By midmorning Monday these levee breaches had effectively sealed the fate of this uniquely vibrant American city. Not that most of us understood this at the time. In fact, few of us did, including me. As it would turn out, this lack of information and communication pretty well defines everything that went wrong in New Orleans over the following weeks, a systemwide breakdown that started immediately on Monday morning. Those of us stationed at the emergency operations center in Baton Rouge knew that Plaquemines and St. Bernard parishes to the east and southeast of New Orleans had to be decimated. Verification unnecessary, really. We knew that the Mississippi shoreline had to have suffered terrible damage. Verification also unnecessary. Archival work now tells us that the National Weather Service office in Slidell posted a flash flood warning at 8:14 A.M. for the Lower Ninth Ward, following breaches in the Industrial Canal levees. No one at the operations center got that message at the time, and then the Slidell office lost all contact with everyone as Katrina passed just to its east. At the center, feeds from the National Data Buoy Center, the U.S. Geological Survey, and both New Orleans airports were all down. We had no storm-surge, rainfall, or wind-speed information. Apparently Coast Guard people knew that there were residents stranded on roofs in the Upper Ninth, that there was a minimum of six feet of water in the Lower Ninth and ten feet in St. Bernard Parish. We had been predicting flooding in those neighborhoods, but nothing like that. Why didn't we know?

Everyone appears to have been working in a vacuum, totally reliant on information gleaned from the various briefings and, most important, word of mouth. The emergency operations center was still fully functional. We had power, Internet, you name it—but no communication out of the hurricane-damaged areas. Everyone everywhere appears to have been working in the same vacuum. Senator Mary Landrieu would admit soon that communications were "entirely dysfunctional." So much for the state's Smart Zone system, which had totally failed to cope. A lot of people are under the impression that cell phones are somehow above it all, but they're not. The system requires both electricity and functioning landlines. Any individual or group relying on them was in for a major surprise. Simply put, along with everything else during Hurricane Katrina, we had a ridiculous, tragic failure to communicate. Several times I thought about the ease with which I'd stayed in contact with my sailing friends crossing the North Atlantic earlier that summer.

By midafternoon Monday the storm was over, long gone, leaving behind partly cloudy skies and mildly gusty winds—a beautiful afternoon—but with water pouring into most of New Orleans through the multiple levee breaches. Why, then, did the phone call that night about the rising water in the nursing home puzzle me so much? (I related that fresh or salty story in the first paragraphs of the Introduction.) Governor Blanco did mention breaches to NBC early in the morning, but she could not have known about the canal breaches, which hadn't happened yet. Perhaps she was referring to those earliest breaches in St. Bernard Parish, or perhaps she was equating overtopping with breaching, as many would.

During hurricane emergencies we man our LSU booth around the clock, with each shift headed by a very competent individual from the Southern Regional Climate Center. From 2:00 A.M. through midafternoon Monday this was climatologist Barry Keim. He heard nothing about any breaches the whole time, and the LSU booth is right opposite the Corps's booth. Look up from one of our computers and you look a Corps employee right in the eye. As I've related, when the report of flooding at the nursing home came in to our LSU cubicle Monday night, I am sure the Corps of Engineers employee heard it, along with me and others, and didn't say a word.

I arrived at the EOC just as Barry was leaving midafternoon, and he expressed his belief that apparently things weren't as bad as they could have been. Later I chatted with Marc Levitan, and someone mentioned "dodging the bullet." All the while, some FEMA people knew about the devastating flood under way. FEMA photographs dated August 29 have since been posted on the Web. Marty Bahamonde, a twelve-year veteran and apparently the only FEMA staffer to ride out the storm in New Orleans, told the Senate investigating committee that he had informed one of Michael Brown's assistants of the catastrophic flooding at 11:00 A.M. Monday. Brown in turn told the same Senate committee that an e-mail went to Chertoff at 9:27 P.M. and that the White House was informed before midnight. Yet Michael Chertoff told a House investigating committee that his last report that night had been a thumbs-up. "It appeared that the worst was over," he stated. Unbelievable.

People at the parish emergency operations center at city hall in New Orleans (a smaller facility than the state EOC in Baton Rouge) knew about the 17th Street Canal breach at 11:00 A.M. Colonel Richard Wagenaar of the Army Corps of Engineers would tell PBS's *Nova* that he knew by midday Monday about the inexplicably high water in the middle of New Orleans, because he was blocked by it on his first tour of the city. He said his group turned around and went back to their bunker and would reassess the situation the next day! Al Naomi, a project manager for the Corps, is quoted in *CNN Reports—Katrina State of Emergency* to the effect that his people had confirmed the 17th Street Canal breach by 2:00 P.M. Naomi said the news "was disseminated. It went to our [EOC people] in Baton Rouge, to the state, FEMA, and the Corps. The people in the field knew it, the people here [Corps offices in Louisiana and Mississippi]."

I read those statements with astonishment and can only conclude that something went very, very wrong in both FEMA's and the Corps's communications pipeline. Some of their people knew about the flooding, yet the word apparently failed to get out to the other agencies at the main emergency operations center. Or if the news did reach their people at the EOC in Baton Rouge, it wasn't passed around the room. This is the most devastating fact to me, because two thirds of the drowning deaths in New Orleans Metro bowl were due to the flooding from the three breaches on the drainage canals. That's the conclusion drawn from the most comprehensive mortality figures on the tragedy, data gathered by one of our hurricane center researchers as part of our GIS program. How many of those lives could have been saved if everyone who knew the facts had done everything in their power to make them known as widely as possible? Granted, spreading the word wouldn't have been easy on Monday. Power was down throughout the region. But both WWL-TV and WWL-AM never went off the air. Couldn't those who knew about the breaches have gotten immediate word to their counterparts at the operations center?

I'll put it this way: Knowledge of those breaches seems to have been tightly held, and the failure of the word to spread—whether due to incompetence or oversight or terrible communications—figured directly in the tragedy to come.

In February, House Republicans issued a blistering report about the delayed evacuation and the failure of the Bush administration to act on the early reports of levee failures. The day Katrina hit, between twelve and fourteen hours after most of the breaches occurred, Secretary Chertoff and the White House had been informed. This bold denunciation was titled "A Failure of Initiative." I rest my case!

FIVE

THE SECOND-WORST-CASE SCENARIO

On Monday morning Brigadier General Brod Veillon and Colonel Pat Santos were on duty at Jackson Barracks, the headquarters for the local detachment of the Louisiana National Guard. The barracks is in the Arabi neighborhood, in the St. Bernard Bowl on the eastern side of New Orleans. Santos had grown up at the barracks with his parents and had seen the major flooding from Hurricane Betsy in 1965, when the waters didn't even reach the bottom of the steps of the headquarters building. No surprise. This is relatively high ground near the Mississippi River. It doesn't flood. But early on this morning forty years later, Santos and Veillon and the other guards on duty watched in wonder as the water rose from nowhere to the first-floor windows in less than an hour. This building is two miles from the Industrial Canal breaches to the west. Such was the volume of water pouring through those gaping holes. I first heard the story of what happened next in November, when a group of us from LSU led a contingent of staffers from the House Committee on Homeland Security on a tour of the breaches across the city. I then received a note with the same story from a friend who used to work at LSU, whose husband is the commander of a National Guard unit that was stationed at Jackson Barracks on Monday.

In Louisiana the Department of Wildlife and Fisheries (DWF) has primary responsibility for search and rescue during floods, because they have the boats. On Sunday morning, six large boats and a dozen smaller ones were prepositioned at the barracks, because department officials had attended all of the surge-model briefings at the EOC and knew what was coming. On the other hand, they had no idea how high the water would reach, because the levee failures had not been anticipated. The boats were positioned on trailers, so as the water rose the boats did likewise. The DWF people and the guards had to use steel cutters on the tie-downs in order to release the craft. Two guards were assigned to each boat, and after the winds had died down they were on the water by 3:30 P.M. Monday. By that time the water was going down. (The water line clearly visible months later on these houses in the Lower Ninth, four feet up the front door, say, did not indicate the highest level the water had reached in the initial deluge Monday morning. Debris on the roof would tell us that the water was at least that high, an important factor in the forensics investigation. Another important determination was the highest storm surge, not counting waves. This water line could only be found in protected spaces such as closets.)

Overall, that completely unexpected and unprecedented flood at the barracks played hell with the first-response capability of the local National Guard units, but this remnant and their DWF counterparts were still the first rescuers on the water anywhere in the city, I believe (though I can't be absolutely certain of this). These men and women fanning out from Jackson Barracks could not believe what they confronted: In every direction, hundreds of bewildered, bereft people and their pets on porches, if they were lucky, or rooftops, more likely. The boaters set to work, ferrying the rescued to the courthouse of St. Bernard Parish, to the levees, to any high ground. Just before dark, Coast Guard and National Guard helicopters started to move those on high ground to the Superdome, therefore setting the stage, unwittingly, for the debacle that was to follow at that building over the subsequent few days. The boats had to cease operations with darkness, because the overhead power lines were now right at eye level for anyone on the water. (Some of the helicopters could operate at night with spotlights and infrared sensors.)

So there was a bit of rescue work already underway on Monday, but of course almost no one knew about this, either. By unofficial count, these small teams—many of whom had lived nearby and lost their own homes—pulled two thousand people from jeopardy in the days to come.

At 3:00 A.M. Tuesday morning a BBC producer called my cell phone, anxious for the latest. (In and around Baton Rouge we did enjoy spotty cell coverage.) My wife loved that, but she soon got used to it. Once your phone number gets into circulation, it's all over. Maybe if I had "slammed down the receiver" for those first calls in the middle of the night a different word would have begun to spread, but I didn't think that was part of my basic job description, which for four years had been to understand what such a storm would mean for this part of the country, aid in preparation, and spread the word. Now, when the worst was actually happening, was no time to back off.

As for the latest, well, early that morning—now twenty hours after the catastrophic flooding had begun in New Orleans—the world at large, including me, was *still* under the misimpression that the famed city had dodged the proverbial bullet. As I write this I'm looking at a congratulatory e-mail from a well-informed friend who works for a major oil and gas company in Maryland, time-stamped 6:22 A.M. Even CNN was confused: Later that morning it reported a "new" levee failure. Maybe that was a failure newly discovered by the network, but it was not a new failure. I couldn't help the BBC producer, who couldn't help me. Like a lot of people I learned the fate of New Orleans when I tuned in to WWL-AM driving to Baton Rouge that morning, as I described in the introduction. Then I knew what the reports about "new flooding" meant. They meant that the water in the nursing home had been salty. They meant that my worst fears were coming true. Yet at a press conference at four o'clock Tuesday afternoon Senator David Vitter said about the water, "In the metropolitan area in general, in the huge majority of areas, it's not rising at all. . . . I don't want to alarm anyone that New Orleans is filling up like a bowl."

At that moment the water was still rising in the Orleans Metro

Bowl, thanks to the levee breaches on the 17th Street and London Avenue canals, was stable in the East Orleans Bowl, and was falling in the St. Bernard Bowl (but still very high nevertheless). So I definitely wanted to alarm everyone that New Orleans either *was* filling up like a bowl or had *already* filled up like a bowl. We no longer needed to tell the flooded residents to run or swim or climb for their lives, because they now had no choice, but every official, bureaucrat, and functionary needed to understand what was happening and to get moving. Apparently the message wasn't getting through. A FEMA spokesperson agreed with Senator Vitter about the bowl not filling up, and added, "That's just not happening." His boss, Michael Brown, dispatched one thousand employees to the stricken area—and gave them two days to arrive. Brown also reminded all first responders to be sure to check in first with state and local officials. By all means, check in.

Most of us with the LSU Hurricane Center took a different approach with our interviews. If anything, we embellished the extent of the flooding. For one thing, we didn't yet know where it would end or at what depths. For another, all remaining residents needed to get out of New Orleans. (Mayor Nagin issued his bulletin to that effect the following morning.) In retrospect, it seems to me that it wasn't until midafternoon Tuesday that the terrible implications of the levee breaches became indisputable. We saw the pictures of people stranded on roofs, on overpasses, on levees. We saw the tsunamilike damage in southeastern Louisiana and the Mississippi coastline. We saw the early looting footage. We saw the rapidly deteriorating conditions at the Superdome, suddenly besieged with people flooded out of their homes and brought in by rescue helicopters, but with those same floodwaters now closing in on the big building itself. Staffs were evacuating patients from hospitals to the Causeway Cloverleaf overpass, where ambulances could then pick up these patients and drive them west on Interstate 10. So faint glimmers of the chaos to come were getting out, thanks almost entirely to our often maligned media, which did an absolutely amazing job during Katrina, in my view. On Monday they had been way ahead of everyone else. On Tuesday they kept their lead. For the next week the reporters and the cameras got to places that FEMA never reached, or so it seemed.

Invisible on Monday, the Army Corps of Engineers now surfaced on Tuesday to take a look at the problem. They arrived at the intersection of the Old Hammond Highway and the 17th Street Canal in tandem with officials from the state Department of Transportation and Development (DOTD), who were touring as much of the city as they could reach, trying to pinpoint the sources of the flooding and assess the status of the pump stations. Standing on the bridge over the canal, these stunned representatives looked at the gaping breach on the east side of the levee, a couple of hundred yards wide, a couple of hundred yards away. Twenty-four hours after the levee failure here water was still pouring into the city, with every home to the east deep underwater. To the west, just one hundred yards away on the unbroken side of the canal, the city was dry. For the most part, therefore, this canal marked the western edge of the major flooding. Soon officials from the Orleans Levee Board joined the group, but not by car. Their whole territory was flooding, so they had walked along the top of the flood wall from the district's storage yard behind the Lake Marina and Yacht Club, about half a mile away.

About 11:00 A.M., a few helicopters appeared overhead—a unit of the Texas Air National Guard. One of the choppers found a place to land, and the pilot conferred with DOTD's Mike Stack about hauling in sandbags from a nearby staging ground set up earlier by the levee board. Over the next couple of hours these choppers dropped maybe ten bags, then flew off across Lake Pontchartrain to refuel and never came back, no doubt rerouted to rescue efforts, because the dire straits of those remaining in the flooded neighborhoods was now known to all.

By the letter of the law the Corps was not responsible for responding to that breach at the 17th Street Canal—the Orleans Levee Board was—but that organization was effectively underwater and out of commission. The DOTD folks were operating under the general assumption that the Corps would take control of the situation. In fact, however, the Corps's personnel at the site were ordered to leave when the sandbag drop started on Tuesday. They did nothing at the site on Tuesday. The Corps would return the following day on a consulting basis, then arrive en masse and take over the job on Friday. Is it conspiratorial of me to imagine that their actions might have related in some way to the arrival Friday afternoon of President Bush to view the now world famous breach?

On Tuesday afternoon Mayor Nagin said that the city had dodged the worst-case scenario on Monday, when Katrina struck, but was now facing the second-worst-case scenario. Fairly phrased, I think. With a stricken sadness in his face and eyes, the mayor also acknowledged the harsh truth: The water would stop rising throughout the city only when it equalized with the water in the lake. From the air most of New Orleans looked like a houseboat community. Someone said, "We are *below* ground zero." For us at LSU the immediately pressing question was how far below? To figure that out we needed to know how high Lake Pontchartrain was, and the water level gauges at the midlake site were no longer transmitting data. So I headed out, with an ulterior motive as well. I wanted to check on my sailboat. I rushed home, picked up Lorie and the inflatable dinghy, and headed for Madisonville, talking to reporters via the cell phone until I lost coverage. Once off the highway we encountered lots of downed trees and power lines. At Madisonville the police had set up a checkpoint, but that's why I have my bright yellow LSU hurricane center sign on the dashboard. It always works, I'm proud to say. Once on the water we headed for a spot where I knew I could get the lake's water elevation, or at least a good estimate, and also determine if the Tchefuncte River was still rising, which was quite possible, because some of these river basins had taken in a foot of rain. Then, unbelievably, the cooling water pump on my dingy outboard engine quit. In eighteen years, this engine had never given me a problem. I paddled ashore, packed up the dinghy, and got across the bridge on the Tchefuncte to Marina Del Ray, where I had another rough water gauge. But I didn't really need it, because we had to wade in chest-deep water to get to the dock site. This water was about six feet above sea level, with the mean at that locale two feet. And the water was still rising. Not good for New Orleans.

Without the dinghy we now had no way to get to our hurricane hole, but just as we were about to leave in disappointment, the marina owner's son offered to take us up the river. Plenty of trepidation on the way, but to our joy the *Maggie* was fine, for the most part. But she was now awfully close to the bank, and when we pulled up on the anchor lines to move her, we couldn't. Something had rolled in over the lines, and all were stuck. Looking more closely at the anchorage, I could see what had happened. Sometime after I'd added anchors and lines on Sunday, a houseboat had pulled in and moved one of my main anchors. This just isn't done, not by real boaters. In my book, you respect other people's property. This refuge had lots of space; there had been no need to interfere with my carefully designed anchor spread. As it turned out, that houseboat owner cost me time and some tough conversations with my insurance company. On the other hand, it could have been worse. I had feared the worst for my boat.

On the river we saw a massive fish kill and schools of small minnows swimming close to the surface, apparently sucking air, a sure indication of low oxygen levels in the water. I wondered aloud what would happen to the whole lake system once the pumps in New Orleans started to empty the "witch's brew" out of the city and into this water. At the first chance I cell-phoned the information to the LSU desk at the state EOC. I also phoned surge warrior and GIS genius Ahmet Binselam and asked him to pull up the LiDAR data set for New Orleans. (LiDAR stands for light detection and ranging, a system for projecting laser beams from an aircraft and working with the beams reflected back to the source. [Radar is an equivalent tool using radio waves.] This technology has given us an extremely accurate elevation map for New Orleans—a dry New Orleans.) Combining this elevation map with the elevation of the floodwater gives us the depth of the water at any chosen location, and I asked Ahmet to generate three images for our Web page, one each to show the extent of the flooding if the water in the city stabilized at sea level plus five feet, plus four feet, and plus three feet. At the time, the lake was at plus six, maybe even more, but it had to start falling fairly soon. How far and how fast I didn't know, however, so it made sense to cover all bets in the likely range. This would give the emergency responders an idea of just how severe the flooding would ultimately be. We could also send all of the reporters to the page, all of whom wanted to know how deep the water would be. They also wanted to know how long before stabilization. I thought it would be a few days.

No sooner had I returned to my office at LSU than I got an e-mail from Mark Schleifstein, who was "holed up" in his "moat-surrounded newspaper building." While some of the other media were still reporting that the worst was over, with New Orleans spared, Mark knew better. He could see the waters rising at the *Times-Picayune* building about a mile north of the Superdome. He also heard about fellow staffers whose homes in nearby Lakeview were being flooded, primarily by water from the London Avenue breach. (A few hours later the editors of Mark's newspaper would have to order the evacuation of their building and head for Baton Rouge. From borrowed offices at LSU, they put out their newspaper through what became the vitally important Nola.com Web site.) I wasn't surprised that the coauthor of the "Washing Away" series instinctively understood that the critical issue regarding the flooding would be the level of the lake over the upcoming days. Could I give him any help? A few hours later, yes, as Ahmet posted the first map for flooding in New Orleans. What a wonderful person Ahmet is. He went many days with no sleep and was always pleasant. He had been severely beaten up the previous Halloween, possibly because he is a Muslim, and now he has a steel plate in his skull. The responsible parties are still unknown. Where are they now? I want them to know what a mild-mannered, peaceful person their victim is, and what a huge role he played in saving lives during Katrina.

That Tuesday night on CNN Larry King put out the number that crystallized the catastrophe for everyone: Eighty percent of New Orleans had flooded, with the waters still rising. Governor Barbour described the utter devastation along the Mississippi coastline, graphically confirmed by the footage. There were estimates of five thousand buildings destroyed in Biloxi, one fifth of the city. The storm had no respect for property values, that's for sure. On the opposite end of the economic scale from sections of the Lower Ninth Ward in New Orleans were the couple of hundred half-million-dollar homes in the South Diamondhead development on Bay St. Louis and the older homes on the water in Pass Christian, all swept away.

At the end of the program it was my turn, by telephone. Since I was the public health expert on the show, King wanted to know about West Nile virus. As it happened, I'd opened an e-mail on that subject that morning from a colleague at the University of Texas, Brownsville. I replied to King that "we're going to have a whole lot more folks gone down with West Nile." That prediction proved right. In December 2005, the Centers for Disease Control reported that the number of cases in Louisiana, Mississippi, Alabama, and Texas had increased by about 24 percent from 2004, all due to Katrina and Rita.

Our host asked me about toxic mold.

Definitely a problem for anyone wanting to restore a house that had stood in water for weeks.

"What's your read on the future?"

"It's very bleak." I then amplified on this assessment.

"How long is this going to take to rebuild?"

"Years. Years and years and years." I then amplified on this assessment.

King asked Governor Blanco if she wanted to comment on my list of issues. "They're all our worries," she said. "It's the very list of things I've been worrying about all day long . . . how to make sense of all this. A million people homeless is not something that happens just every day." The governor took some criticism in some quarters for looking, in those first few days after Katrina hit, like a deer caught in the headlights. I think that's unfair. To my mind she and Mayor Nagin have a much better defense for their responses immediately after the storm—set aside planning—because they were inside the story. It had happened to them. The situation in Louisiana and Mississippi now really was, as King replied to the governor,

"Unfathomable." The enormities of this disaster were just starting to settle in, and for many it was totally unbelievable. Everywhere you looked the wheels were coming off. Even a hardened professional such as Walter Maestri, head of emergency management for Jefferson Parish, burst into tears over the radio as he pleaded for help.

At home that night, with the water in much of the city still rising, I got a wonderful e-mail from a student in my DSM 2000 class the previous spring (an introductory disaster science and management course). John Tanory had a seventeen-foot boat with a 140-hp Evinrude motor in Houma near the coast, and he and his brother wanted to help. I suggested he contact the Jefferson Parish EOC, and wished him luck. I wonder whether John and his brother headed for New Orleans, early volunteers in our "Operation Dunkirk"—on the home front this time, and sixty years after the first one. The main staging ground for the freelance boaters working in the Orleans Metro Bowl became the corner of Poydras and Loyola streets, two blocks from the renamed Sewerdome. Rescuers could tow their boats right up to this intersection from the west, launch, and get to work saving lives. Over the following weeks, then months, I drove by this intersection often, and long after the city was completely dry there were still a couple of boats tilted on their sides here, looking exhausted.

SIX

INTO THE BREACH

Everyone finally knew the full extent of the flooding—Larry King and others had latched onto the "80 percent" estimate (I used 85 percent)—but on Wednesday morning, when we at LSU actually saw the first QuickBird satellite images supplied by Digital Globe, we were stunned. If anything, even my estimate seemed too low regarding the Orleans East, Orleans Metro, and St. Bernard bowls, almost all of which were forbiddingly dark and murky in these images. Altogether, 148 square miles of urban flooding. Had any large city anywhere ever been so devastated by water? The Dutch experienced massive flooding in 1953, but not in any of their very large cities. None of the Mississippi River floods had inundated major cities. Florence was famously flooded in 1966, but not to this extent. All I know for certain is that no such scene had ever been captured with such compelling images.

You have to understand. All of us at the hurricane center—and in other offices around the state and the country, for that matter—were almost as familiar with the basic scene on the table in front of us—New Orleans, uniquely defined by Lake Pontchartrain and the big river with the large crescent that gives the city its nickname—as we were with pictures of our own families. In a way, we were looking at our professional lives. It was quite strange, and these first pictures had a sobering impact on all of us. We could now see the enormity of the problem facing everyone still in the city and everyone trying to help everyone still in the city. Prone to worrying about worst-case scenarios anyway, I found this imagery really depressing—morbidly so. I felt a real sense of dread.

The first batch of ten pictures included three before and after comparisons, one each for the Industrial, 17th Street, and London Avenue canals. Looking at them again months later, I now realize that the one for London Avenue includes only the breach on the west side near Robert E. Lee Boulevard, not the one on the east side, near Mirabeau Bridge. Even though this breach was less than half a mile from the other one, no one had officially identified it! Not until Sunday—six days after the storm—when I joined a flight over the region, could I confirm to state officials the *two* breaches on the London Avenue Canal. That's how shaky everyone's information was that first week after Katrina swept through. It's unbelievable, considering all of the resources theoretically at hand.

Still, at the two hurricane centers we did have a tremendous amount of background data and information in the Geographical Information System (GIS) database, and during that first week, especially, we hustled to get it out there to everyone who asked for it—FEMA, the Centers for Disease Control, the Department of Health and Human Services, the state police, the Department of Wildlife and Fisheries, the state Department of Health and Hospitals, the emergency operations center—and some who didn't ask, because they didn't know we had what they needed. Even some of the state agencies were surprised by the cornucopia at our fingertips. Late Tuesday night, when it was already obvious that our computing power wouldn't be able to handle the demand for the GIS products and the maps, Kate Streva, Ahmet Binselam, and Hampton Peele, old friend and GIS wiz with the Louisiana Geological Survey, hauled all the machines from the offices at LSU to the cubicles at the EOC and set up a twenty-four-hour-a-day operation. Kate, third-trimester pregnant, organized the whole thing. When I was finally able to persuade her to take that baby home, John Pine of the geography department, and also a member of our team, took over with a flourish. On Wednesday afternoon I was at a briefing at which a state employee gave his update using a small-scale road map taped to the wall. This was the kind of map they hand out to tourists! Wait a minute, I thought. Given the national exposure of these briefings, given our incredible cartography resources, we can and will do better than this. I rushed back to LSU and called on John Snead, a good friend, head of the cartographic section of the Louisiana Geological Survey and part of our GIS team. John produces the official state map; I asked him to produce a large-scale version depicting southeastern Louisiana. The next day we taped this superior map to the wall in the press briefing room at the emergency center. Then everybody wanted one, which was fine. John's team ran off dozens of copies. Clearly, already, Katrina was shaping up as a major catastrophe, and in order to properly articulate what was happening where, everyone involved needed good maps. This is one of the unwritten laws of disaster management.

The Board of Regents authorized us to go to "full operational support"; that is, our researchers were authorized to participate in any support activity related to their expertise. On Wednesday night we had a big meeting to coordinate all this. Nor were the two hurricane centers the only disaster resources on the campus. Not at all. There are also researchers at Louisiana Geological Survey and in the geography, environmental studies, civil engineering, and other departments, and everyone wanted to share everything with everybody. That's the simplest way to summarize the response across the board in the Katrina emergency, and I don't exaggerate. GIS-savvy people I had never met suddenly stepped forth with their expertise and enthusiasm. In this respect, those first weeks were gratifying. I've already noted with emphasis that pure science is sexier than applied science for most universities. Even with all the visibility our LSU centers have received since they were founded in 2000 and 2001, respectively—and not just visibility, but with good, useful research—even though we bring in large sums of money, we have always been shortchanged when it comes to resources. Do I sound peevish? So be it. I think the blinders that upper administrators of many universities utilize in defining their institutions' role in society hurt everyone, and they don't endear the universities to legislators either, especially not in Louisiana. Anyway, given the second-class status of us soft-money scientists at LSU, there was some satisfaction during

Katrina when we could not only say "We told you so!" but also contribute to the immediate solutions as well.

Early on Thursday morning the water in the Orleans Metro Bowl finally achieved equilibrium with Lake Pontchartrain at three feet above mean sea level. In the St. Bernard Bowl to the east the water level dropped from its highest mark of plus-twelve until Saturday, when it was also stabilized at about plus-three, while in Orleans East the water level never got as high as sea level (though these neighborhoods flooded terribly, because some are minus-ten in elevation). I described in the previous chapter how GIS whiz Ahmet Binselam had combined our LiDAR-based elevation maps of New Orleans with three different hypothetical floodwater depths—three, four, and five feet above sea level—to show everyone where the waters would be after equilibrium. He posted those models on Tuesday. Over the next week all of the agencies engaged in search and rescue, including FEMA, the Coast Guard, the National Guard, Louisiana Wildlife and Fisheries, and the state police wanted the plus-three map to guide their efforts. Because resources were so limited, extremely jeopardized sites such as nursing homes were searched for survivors first, and those sites were prioritized according to water depth. Generally speaking, everyone except some of the freelancing rescuers tried to proceed from the deepest to the shallowest areas.

On Thursday we had a bit of a dustup with the French company that produces and owns the best high-resolution satellite imagery with which to determine flooding on the commercial market. LSU's Earth Scan Lab had received some of this SPOT imagery from the remote sensing laboratory at the University of Miami and immediately posted it on various LSU Web sites. The value of this imagery over the QuickBird photographs we had gotten on Wednesday morning is that it is very easy to discern flooded versus nonflooded areas with SPOT. Additionally, the resolution is good enough that I could use it to "inspect" the levee systems for the whole region to see just where and how bad the breaches had been. For example, at this still early stage of the disaster, we didn't have solid information about conditions in the lower St. Bernard Parish east of New Orleans. Anecdotal reports were just about it, and they seemed almost unbelievable. But by using the SPOT imagery to find all the levee breaches we were able to see that the anecdotal was all too real. About four miles of those levees had been completely destroyed. Anyone who hadn't evacuated would have faced the almost unhindered storm surge and water as high as fifteen feet above sea level! If still alive, they would be in desperate straits now. So this SPOT imagery had proved its value almost immediately, but no sooner had we started using it than the folks in Miami advised us to cease and desist, to take it down from all LSU Web sites, and to advise anyone who had already accessed the data—six hundred users, we quickly counted—to also cease and desist. Our friends in Miami said the images were strictly NDA—No Distribution Allowed. So said the French company that owns the SPOT imagery.

I quickly found out that this company already had a reputation in Louisiana for aggressively defending its copyrights. Some time earlier the Department of Environmental Quality had purchased SPOT panchromatic imagery for the whole state, fused it with Thematic Mapper imagery, and released the beautiful package of images and information to the public. The company immediately chastised the agency and threatened a lawsuit. When it persisted with this threat, the agency complied and has never done business with that outfit again. On Thursday morning I e-mailed our contact in Miami the following note: "As I write this e-mail people in New Orleans are dying. The rescue squads as well as the public health and other emergency medical personnel really need this data. This is a time of national and international crisis. Tonight I will be a guest once again on the Larry King show and I will make a point of telling the world about this heartless approach to the New Orleans disaster, Ivor."

I was really angry. It was my pleasure to threaten that company with what I sometimes get the impression is just about the only thing any big company fears: bad publicity. (I don't know about in France, but in the United States I understand that companies also fear juries, which is why they pour so much money into the so-called tort reform movement.) In any event, my little threat worked like a charm. In not much over an hour our colleagues in Miami told us that they had permission to set up a site to use the SPOT imagery of New Orleans for "humanitarian purposes."

Before the taping of my remarks for Larry King's show Thursday night, I ran into Senator Mary Landrieu, with whom I'd shared the King platform the night before and who was preparing for an interview with another network. I told the senator I was pretty distressed. I was pretty sure that King would ask my opinion about the response, whether "the wheels had come off" or something like that, and I felt the wheels of the emergency response *had* come off, and I would probably say so. She suggested a softer approach. Mindful of my status as a lowly soft-money scientist at LSU, and of the general wisdom of avoiding the overtly political, I was swayed by her advice. Holding back wasn't easy, but I did it. I told King, "The wheels haven't come off, but the wheel nuts have come off." Maybe that was a bit too technical for the broad audience, but I'll stand by the assessment now. Then I added (as I stated in the Introduction), "We in Louisiana can only trust that our governor, and especially the president, are putting all the resources of the federal government and our mighty military to bear on this problem."

Anderson Cooper was also on the show. Earlier that day he had famously come close to blowing his editorial objectivity while reporting from Biloxi, Mississippi, I think, although he'd already been all over the region. He was just tired of hearing politicians say they understood people's frustration. Had he heard Governor Haley Barbour's previous remark to that effect on this very show? I don't know, but he said, "It's not that people are *frustrated.* It's that they're *dying.*" Anderson explained that some people on the streets could pick up news from satellite radio, and they heard the politicians congratulating each other about their great relief work, and it was driving them crazy. Actually, it was driving all of us crazy. As any disaster science specialist knows, the presence of politicians glad-handing and congratulating each other in the response phase of a disaster can only mean that things are going badly. All you had to do was watch the TV during the first days of the Katrina story to know the wheels had come off! I heard Anderson's interview while I was waiting my turn, and I really respected him for speaking out. A few days later I spent a few hours with him, and I told him that I admired him for saying what needed to be said. When he said in turn that he thought of me as one of the only honest, sane, and knowledgeable voices out there—well, that really put a lift in my step.

Also on Thursday—maybe Friday, I'm not sure, but right when the finger-pointing started—I found myself in a radio interview with someone who turned out to be a rabid reactionary on the West Coast who right off the bat started hassling me about all the corruption in Louisiana. So what? I said. There has been and is corruption in every state, and even though people down here enjoy talking about it as part of our political culture, I don't know if it's really worse than anywhere else. Chicago? Seems I've heard a few stories about the former Mayor Daley. New York? Boss Tweed. Boston? I understand the Irish had an iron grip in the old days. Los Angeles? The movie *Chinatown* was based on hard facts: The kingpins in Los Angeles stole that water. So how can you throw stones from one thousand miles away? What does any of this matter now, anyway? Any corruption in Louisiana isn't the reason FEMA, at the federal level, was totally unprepared for this catastrophe. This guy really pissed me off. Then I had another interview like that with some Canadian clown. A couple of people who stumbled across that show said I did a good job. I sure hope so. I had no patience for any of that, and I still don't. Where were the planning and the preparation for an event everyone in authority knew would happen eventually? And what are we going to do now?

The whole world saw the encampment beneath the I-10 overpass in Jefferson Parish, where there were no supplies or facilities of any sort, but where the helicopters were bringing in more people all the time. Also in Jefferson Parish was the large group of stranded folks who were directed to the bridge over the Mississippi River for impending evacuation and were turned away by armed cops. On the September 9 edition of NPR's *This American Life*, producer Alex Blumberg interviewed two paramedics from San Francisco who had been visiting New Orleans for a convention. After the storm they tried to escape the city in a number of ways. On foot they were told by police, at gunpoint, to turn back. What was going on out there?

Everyone heard about the woman who had used a door to float her husband's body to Charity Hospital, where the staff was hand-pumping ventilators for patients in intensive care, who were finally evacuated by boat to dry land, where helicopters picked them up. The makeshift morgue at Charity was now in one (or more) of the stairwells, because the actual morgue had been in the basement, which was now underwater, of course. All together there were about ten thousand patients and staff in the marooned hospitals of New Orleans, and these folks deserve—and will probably receive—a book about their remarkable stories.

As the reports piled up on Thursday, just about everyone lost it, I think. The ineptitude and the tragedy had become too much to overlook or explain away. Major General Harold Cross, the adjutant general of Mississippi, complained that he was communicating with his forces by courier, "like the War of 1812." That's truly embarrassing. In any ranking of the early snafus, the patchy communication should be high on the list. Number one, Governor Blanco said. "Interoperability" is the official lingo, and its absence had also been the problem on September 11, 2001, when New York City police officers and firefighters couldn't communicate with one another as they headed into the doomed Twin Towers. Their systems weren't compatible—that is, interoperable—and this failure almost certainly cost many lives. Chastened, disaster officials at every level of government in every state vowed improvement, but Terry Ebbert, the Homeland Security chief for New Orleans, said after Katrina, "From an interoperability perspective, we are worse today than we were before the storm."

How in the world could this be? The explanations are conflicting and complex and ridiculous—for example, the new federally funded 700-MHz radios in St. Bernard and Plaquemines parishes were deaf to the 800-MHz equipment in New Orleans and Jefferson parishes. In the city, a standoff between private-sector vendors vying to provide new equipment had stalled the interoperability initiative. And these are problems *everywhere.* A *Frontline* broadcast in November

2005 identified a grand total of three states that have installed effectively interoperable systems in all jurisdictions: Delaware, Michigan, and North Carolina. Speaking to the problem, Tom Ridge, former governor of Pennsylvania and former secretary of the Department of Homeland Security, blamed the states, because the federal government can't enforce standards. "Bull," replied Warren Rudman, former Republican senator from New Hampshire. The federal government imposes standards all the time, across the board, Rudman said.

The solution here is painfully obvious. We need one interoperable system covering every jurisdiction in this country, because disasters do not respect borders. New York City is on a border. Philadelphia. Portland. Washington, D.C.! If the fifty states and the federal government can't get it together to accomplish national interoperability, shame on them, shame on us.

In the preceding chapter I introduced the scene at the levee breach on the 17th Street Canal. The Corps had left the site on Tuesday—the rudimentary, woefully ineffective sandbag drop wasn't a Corps-authorized job—but the Corps then arrived en masse on Friday, as would the entourage of President Bush. I'm going to relate what happened in the interim in some detail, because the story has not been told before to my knowledge, and it illustrates in microcosm the heroic efforts and ingenuity of thousands of men and women in the weeks immediately following Katrina—and also the profound confusion and what sometimes seemed to be outright obstruction that also marked those days and nights.

The 17th Street Canal is the border between Orleans Parish to the east and Jefferson Parish to the west. Following Katrina it was also the western border of the flood. It was Orleans Parish that was flooding that week, and the Orleans Levee Board was legally responsible for the levee, and therefore for sealing the levee breach, which was on its side of this canal. But this entity had lost all of its equipment, and its employees were staged at the EOC or scattered far and wide. Jefferson Parish was mostly dry, but not entirely, because some of the flooding to its east in Orleans Parish was backing into adjacent neighborhoods to the west, where workers had their hands full throwing together a makeshift levee right about where Interstate 10 crosses the parish boundary. (Without this levee, just a few feet high, a large section of Metairie would have flooded to depths of perhaps five feet.) But there is also the West Jefferson Levee District, which is responsible for the areas of that parish across the Mississippi River. These were totally dry. Most of the "West Jeff" employees were living at the office, because their homes were without services, and in some cases roofs, but at least their homes were not underwater. All of these folks had long-standing experience with breaches in their incomplete flood-protection system—nothing like this breach, but holes and floods nevertheless. Also, levee boards around the state are always helping each other. It is the neighborly thing to do, not to mention the fact that this flooding in New Orleans was an unparalleled catastrophe for the entire region.

On Wednesday morning, therefore, West Jeff officials Harry Cahill and Giuseppe Miserendino, president and deputy director, respectively, arrived at the 17th Street Canal and volunteered their services and their crew to help. Over some kind of phone connection an official with the Orleans Levee Board officially asked the West Jeff team to do what it could. In conjunction with Mike Stack of the state Department of Transportation and Development and everyone else on hand that morning, a plan was devised to build a makeshift road, a working platform, from the bridge to the breach, using the earthen berm as a starting point, even though water was lapping up its side and it was too waterlogged for vehicles. With the right heavy equipment and dump trucks, the job was doable. Once at the breach the crews could either continue building the road all the way across or drive steel-sheet piling and create a dam. Time for that decision later. One point seemed certain: An aerial bombardment of sandbags would not be enough by itself. Someone's quick math guestimated that a squadron of choppers dropping one hundred bags an hour would require forty-eight hours to seal the breach, and even if these numbers were right, where were these helicopters?

How soon could they arrive? The prudent course was to forget the sandbag option, no matter how clean and simple it might seem.

Mike Stack called his department's personnel at the Emergency Operations Center—called them indirectly, that is. Using a landline at the nearby condo of a levee board official, Mike was somehow able to get through to his wife Linda's cell phone. (They had been flooded out of their nearby home and lost everything, and here he was, busting his ass to do this job.) Mike found Linda, she called the EOC, the EOC personnel presumably got in touch with DOTD secretary Johnny Bradberry's office, and in this roundabout way Mike was authorized to start the road-building job Wednesday morning. Corps officials were on hand that morning, listening, "consulting," whatever, but this was not their job. That was understood.

After clearing a ten-mile roundabout route to the breach site from their depot across the Mississippi River (trees, power lines, and debris were scattered everywhere, remember), the convoy of eight or nine pieces of equipment arrived at noon. Meanwhile, Giuseppe and Mike had located rip rap (broken concrete) and crushed asphalt at a couple of stockpiles coincidentally within blocks of the breach, and they "borrowed" some from the lakefront levee itself (which was now in no danger). They also commandeered an excavator from a nearby construction site. Someone could worry about the legal niceties later. The idea was to build a reasonably safe road to provide access to this breach as quickly as humanly possible.

West Jeff had brought along several small units of portable lights, and by 2:00 A.M. that first night the crew had hauled and filled and compacted five hundred feet of workable road along the embankment, with the floodwater right below them, with fewer than a dozen men. These guys knew what they were doing. Because of safety issues—total fatigue—they quit at that hour and returned Thursday morning to continue the job. By that afternoon, the crew had pushed the road close to the breach and were beginning work on a larger platform, a turnaround that could accommodate the equipment that would be needed for the much bigger job of closing the breach itself, by means soon to be determined. (A helicopter also dropped a few sandbags into the breach on Thursday—Giuseppe had brought them from the West Jeff yard—but it was not a very intensive effort.)

This is when everything started to get very complicated. On Wednesday the decision had been made to get a second job going at their work site, using steel-sheet pile in fifty-to-sixty-foot lengths to seal the canal at the Hammond Highway bridge and thereby prevent the lake water from even reaching the breach two hundred yards downstream. To this end, Mike coordinated with Boh Brothers Construction to bring a big crane from Baton Rouge and the steel from Houston. They should be able to start work the following day. When the crew from Boh Brothers arrived Thursday afternoon and began to assess the job of building the temporary dam, traffic congestion suddenly loomed as a potential problem, because the DOTD/West Jeff dump trucks had to cross that bridge with the fill material for the road job. Mike Stack assigned one lane to each crew.

Corps officials told Mike on Thursday that the Boh Brothers crew was now under contract to the Corps, not the state. They asked Mike about getting access to the canal for a Flexifloat barge from which another Corps-contracted crew could work to plug the breach from the water. They agreed that this contractor would get started on that job at 6:00 A.M. Friday morning. The new crew would need only an hour, and they would coordinate with the other crews. That was the plan, but when the DOTD/West Jeff team arrived at seven o'clock Friday, the bridge was totally blocked by the crew preparing to launch the barge. The Corps people told Mike they'd been ordered to do whatever was necessary to get this barge into the water. Mike replied that his crew had to have access to the bridge to build the roadway, and he said he would have people arrested, if necessary. (He now had state police escorts.) Soon enough, the DOTD/West Jeff crew got access to one lane of the bridge and continued building the road to the breach. The barge contractor then asked the Corps for permission to tear down a portion of the flood wall on the Jefferson Parish side of the canal. The levee district's chief of police said he would arrest anyone who damaged those walls without Mike's permission, which was not forthcoming.

Other complications followed. In the end, it was early evening before the barge was in the water.

Midmorning Friday, Corps officials pulled rank and announced to Mike that they had hired their own contractor to take over the road construction and seal the breach from the north end. Testifying before a Senate investigating committee months later, Colonel Richard Wagenaar referred to these events as a "turf war," and he said that an unnamed West Jeff employee "literally blocked our equipment. They would not let the Corps of Engineers operate." Wagenaar admitted that the Corps was not initially engaged, because it was up to the levee district to attempt a repair. So if the DOTD/West Jeff team was succeeding, why did the Corps feel the sudden need to get involved? Harry Cahill, the head of the West Jefferson Levee District, told the same committee that his crew had begun building the road because no other agency was doing anything. He described Wagenaar's depiction as "full of bull."

On Friday, therefore, confusion reigned—chaos, from what I gather—and DOTD secretary Johnny Bradberry huddled with General Don Riley of the Corps and officials from the Orleans Levee Board to get straight, once and for all, who was running this job site. Mike Stack entered the discussions. Giuseppe Miserendino entered the discussions and urged someone to do something about the danger posed to the workers by all the helicopters that had by now joined the action, flying directly overhead with their sandbags not one hundred feet above the ground. When the backwash from one chopper knocked Johnny Bradberry against the flood wall, he told Giuseppe that he'd take care of this, and he did. The choppers thereafter altered their flight paths, for the most part.

Literally hundreds of people were milling around the bridge area on Friday, suddenly including big guys carrying M-16s. This was a puzzle until word quickly spread that President Bush would pay a visit in the afternoon, boots on the ground this time, trying to make political amends for his pitiful flyover two days earlier. His entourage arrived at 3:00 P.M. Governor Blanco, when someone informed her of Mike's lost home, gave Mike a big hug, and then she told the president, and he gave Mike a big hug. (Telling the story, Mike smiles ruefully. The camera has disappeared. The Secret Service asked Mike for his address, for the forwarding of their pictures, but at this writing he hasn't yet received those pictures.)

Did the president's impending arrival have anything to do with the move by the Corps to take over the whole job Friday morning? I don't imagine it was irrelevant, but in any event the Army was now in charge. DOTD secretary Johnny Bradberry said so, and he ordered the DOTD/West Jeff team to work around the clock under the Corps's auspices. The overall idea now was for this crew to finish the road approaching the breach from the north, then work to fill the breach itself from that end, while the Corps used its barges to work on the middle and the southern end of the breach, while also dropping sandbags from helicopters. The dam at the bridge would not be closed off at this time, though it would be three weeks later as Hurricane Rita approached. With the equalization of Lake Pontchartrain and the floodwaters in the city on early Thursday morning, water was no longer flowing in from the lake, so the dam was no longer necessary.

With the Corps now running the operation, the road-building scheme changed. Work on the road actually suffered. It didn't speed up, that's for sure, because the eight trucks employed in the West Jeff system, enjoying ready access to nearby fill material, had maintained full capacity and managed a dump every few minutes. Now there were some long waits between dumps, because the dozens and dozens and *dozens* of trucks hired by the Corps were traveling in convoy to their distant supply source. Moreover, the West Jeff drivers were professionals at this kind of thing. Backing a loaded truck down a narrow, makeshift track above dangerous floodwater—they could handle the hazard, but some of the new drivers hired (for a reported seventy-five dollars an hour) by the Corps had trouble with the task. One of the West Jeff drivers was injured when a piece of debris broke the windshield of his truck. This guy refused medical treatment and finished the job without a windshield, and bleeding from the cuts on his face. These West Jeff levee pros were dedicated to the job.

Late Friday night—4:00 A.M. Saturday morning, actually—one of the subcontractors offered to drive a totally exhausted Giuseppe Miserendino home. On the way this man said he intended to make the new road "tighter." What did this mean exactly? Giuseppe wasn't sure, and he was too tired to find out. A few hours later Mike Stack told Giuseppe that Boh Brothers was taking over the entire operation *and* changing the deployment of equipment. The DOTD/West Jeff crew should stand down and leave just a skeleton crew at the site. Okay, Giuseppe said—he had no choice—but he was pretty certain that the Corps's plan would not work in the extremely confined space on that makeshift road. I'll skip the technical details here. Besides, the change lasted all of ten hours. Late Saturday afternoon the Corps realized that its new system wasn't working as well as the old one. Giuseppe got a phone call to immediately remobilize his men and resume work. With two crews working twelve-hour shifts, the breach was closed Tuesday afternoon. (When the Corps began work at the Robert E. Lee breach at the London Avenue Canal a week later, those crews used the road-building system employed by the West Jeff crew at 17th Street. I wonder whether the Corps completed those new roads as efficiently. I rather doubt it. The one at the 17th Street levee cost sixty thousand dollars, not counting whatever the Corps paid for the material hauled in by its trucks.)

On Saturday Mike Stack was ordered to take the day off and make some kind of living arrangements for himself and his wife. They ended up in a room with another couple at a Motel 6 in Baton Rouge, and they were lucky to have such relatively handy accommodations. Mike first learned about the changes at the 17th Street breach that afternoon from CNN, on which live shots of the work showed the new equipment! Puzzled, Mike was back at the site the following day, by which time the original program had been reinstituted.

Telling me this whole story months later, Mike and Giuseppe couldn't hide the emotions that bubbled up. They didn't try. As their crew had been building the road from the bridge to the breach, firemen and other rescuers were using the same bridge as a drop-off point for the stranded survivors they were plucking from rooftops. One day a fireman asked where he should take an eleven-year-old girl whose parents were gone. No one had any idea what to do with this girl. Where is she now? Mike shakes his head. One guy who had commandeered the second floor of a neighbor's house boated up to the bridge daily, trudged off in search of supplies, returned, and boated back to his new home. Food was scare. The levee district and local folks from the dry side of the canal brought some meals to the workers, and Mike brought what he could from his new home at the Motel 6. A cache of cookies held everyone in good stead.

Mike's main swamper at the job was Justin Guilbeau, whose nearby apartment was underwater but who nevertheless worked at the breach virtually nonstop for a week. He told me about the day he was in a boat with a fireman, who leaned over and pulled a cupcake in its plastic wrapping from the stinking green slimy water. After inspecting the package for leaks and finding none, he ripped it open and gulped the badly needed meal. He was that hungry.

According to the first bulletins that hit the airwaves, cops had just shot and killed four construction workers at the 17th Street breach. No, wait a minute, that wasn't right. Cops had shot and killed four *snipers* who'd been shooting *at* the construction workers at the 17th Street breach. That sounded more like it, but no, that wasn't right, either. Maybe the cops had shot and killed snipers over at the Danziger Bridge on the *Industrial* Canal. No, maybe that's not what happened either. In fact, it wasn't. Not even close, but to this day no one is sure what incident started this particular rumor. The same holds for thousands of others. I guess people as well as nature abhor a vacuum, because rumor ran amuck, instantly it seemed, much of it fed by photographs and video footage, which can be highly selective in their emphases, as we all know, and also highly misleading. The media did a great job the first few days, but certain elements sure got into rumor mongering. Looking back, I think the urban myths that took root during the Katrina emergency were just as damaging, in some ways, as the lack of information. Terrible misfortune and rolling incompetence were half of the story that first week. The other half was crime allegedly so pervasive and so violent that "chaos" seemed to be the operative description of the scene.

A visitor from Philadelphia said, "It's downtown Baghdad. I thought this was a sophisticated city. I guess not." Someone at Children's Hospital said that their worst problem was the *rumor* of looters storming the hospital, which was running pretty smoothly on generators. A doctor reported—or was reported to have reported, to be exact—witnessing patient evacuations coming under sniper fire. We read and heard about shots fired at ambulances, at helicopters, at cops, at utility workers, at the rescuers in the boats, at just about every moving human target. The state's DOTD broadcast a message over the airwaves on Wednesday to discourage freelancing boaters from hitting the waters of the city. Too dangerous for them. FEMA suspended boat rescue operations for the same reason. (Knowing FEMA, probably also because these folks weren't signing in and picking up their badges. Still, the boaters were pouring in, and it's a good thing. The *New York Times* ran a story about Guy Williams, president of the Gulf Coast Bank & Trust, paddling his canoe along the streets and finding 170 people stranded in one apartment.)

On Wednesday night Mayor Nagin ordered the fifteen hundred cops on duty to stop their search and rescue and concentrate on law enforcement. He said the looting "started with people running out of food, and you can't really argue with that too much. Then it escalated to this kind of mass chaos where people are taking electronic stuff." Now the bad guys "are starting to get closer to heavily populated areas, hotels, hospitals, and we're going to stop it right now." Later in the week he would blame the crime on drug addicts. Governor Blanco said the National Guard in the city were "locked and loaded." Both ordered mandatory evacuation of the city. "We have to," Nagin said. "It's not living conditions." True enough, but the evacuation never really happened. Joseph Matthews, director of New Orleans's Office of Emergency Preparedness, said, "The city is being run by thugs." Police superintendent Edwin Compass said, "The tourists are walking around there, and as soon as the individuals see them, they're being preyed upon. They are beating, they are raping them in the streets."

In Baton Rouge, no sooner had the city opened an emergency shelter at the River Center, a large convention-type facility, on Tuesday, than the rumors started flying, many of them apparently stemming from a routine dispute at a Chevron gas station. All of a sudden we, along with New Orleans, were surely going to be at the mercy of murderers and robbers. Our mayor, Kip Holden, complained that the state had dumped "New Orleans thugs" on his town and ordered a dawn-to-dusk curfew for all evacuees now in the shelters around the city. On Wednesday, a riot report brought SWAT teams to the River Center and precipitated the evacuation of the municipal building. In Slidell paramedics were prevented from going to work for hours because that town across Lake Pontchartrain from New Orleans was said to be overrun by gangs of thugs. On September 2, the *Army Times* gloated that "combat operations are now underway on the streets to take [New Orleans] back. 'This place is going to look like Little Somalia,' Brigadier General Gary Jones, commander of the Louisiana National Guard's Joint Task Force, told Army Times Friday as hundreds of armed troops under his charge prepared to launch a massive citywide security mission from a staging area outside the Louisiana Superdome. . . . While some fight the insurgency in the city, others carry on with rescue and evacuation operations."

I could cull the written and oral record for literally thousands of such stories and statements. But how much of all this was true? There was definitely looting and more serious crime, but was there an *insurgency*? Did looters force the doctors at Charity to hand over drugs? Did the "thugs" who had set up shop at the convention center really beat back eighty cops trying to restore order? Did pirates commandeer rescue boats for their own criminal purposes? What about the rapes reported at the convention center, at the Superdome, on the streets, everywhere? On the same edition of Larry King's show on which I referred to the wheel nuts coming off the recovery effort, Jesse Jackson suggested that these reports and rumors were probably exaggerated. Was he the only one who had noticed that it was the same five guys hustling out of the same Rite-Aid carrying the same loot, over and over and over, on channel after channel after channel? No, he wasn't, but such is the power of the image and the rumor.

LSU criminologist Ed Shihadeh, who was in demand for interviews, has studied the impact of crime rumors. They can actually be scarier than the real thing. Hearing about crime in the neighborhood creates a more generalized fear than being an actual victim. Writing for the Social Science Research Council, Russell R. Dynes and Havidán Rodríguez of the Disaster Research Center at the University of Delaware concluded, "[T]he images of chaos and anarchy portrayed by the mass media were primarily based on rumors and inaccurate assumptions. Some of these were supported by official statements by elected officials."

I certainly concur. For whatever reasons, government officials really exaggerated the "anarchy" issues. The many wonderful stories of selfless sacrifice seemed to have had a hard time competing. However, the truth did begin to catch up. Police superintendent Edwin Compass, who had made such incendiary remarks during the first week, more or less recanted a few weeks later, in an interview with the *New York Times*. (Compass soon resigned under fire.) It's our good luck that David Benelli, head of the sex crimes unit for the New Orleans police, actually lived at the Superdome for the entire week of its service as a refuge and, with his officers, followed up on every rumor that came along. So they were busy, that's for sure, and ended up making *two* arrests, for attempted sexual assault. Regarding everything else, Benelli told the *New York Times*, "I think it was urban myth." On September 26, the *Times-Picayune* provided a major critique of the crime rumors and quoted the Orleans Parish district attorney pointing out that the four murders in New Orleans in the week following Katrina made it a "typical" week in a city that expects two hundred homicides throughout the year.

In Baton Rouge, Mayor Kip Holden, who had complained on Tuesday about the "New Orleans thugs" dumped in his city, tried the next day to defuse the rumors of "looting, rioting or any similar situation." He said his first remarks applied only to actual thugs, not the many more law-abiding evacuees. By Thursday a grand total of one arrest had been reported among the evacuees. Nevertheless, the line at Jim's Firearms was over three hours long. The general perception of an entire populace in jeopardy led LSU and Arizona State officials to relocate the big football game to Tempe. (On the other hand, none of this kept property prices from skyrocketing. Wealthy evacuees from New Orleans bought homes sight unseen, then turned them for a profit when they did see them. Or were such stories just another sort of urban myth? Perhaps, but prices did go up, and within two weeks of the disaster there were essentially no homes on the market in Baton Rouge.)

The *Los Angeles Times* ran a story in November under the headline DOUBT NOW SURROUNDS ACCOUNT OF SNIPERS AMID NEW ORLEANS CHAOS. The many, many sniping stories had really caught my attention, and that of anyone who studies natural disasters, because sniping during such emergencies is almost unheard of. Write that down. Looting, yes; guys with guns, yes; homicides, a few; but random sniping? Highly, *highly* unlikely, and over the weeks and months that followed, more and more of those stories were determined to be something else, or nothing at all. The closest confirmed account was shots perhaps fired at a police station at the edge of the French Quarter. To my knowledge, that's it, and I have spoken with hundreds of sources, including evacuees, first responders (mainly law enforcement), government officials, reporters, and researchers.

New Orleans was never downtown Baghdad.

It helps a rumor to have something to work with, and in New Orleans, that was race. As early as Wednesday the racial aspect of this catastrophe had become impossible to avoid. Most of the people who had remained in New Orleans were black, and they were now either stranded in unimaginable misfortune or stealing everything in sight, or both. That's exactly what the pictures told us. For anyone with any knowledge of the city, surprise at this rather damning image could only have been feigned. Seventy percent of the population of the city is—or was—black. The Lower Ninth Ward, now completely underwater and its population now a diaspora, was 98 percent black. Of the one hundred thousand-plus residents without a car in the immediate family, 90 percent were black. It was not long before some black leaders in New Orleans and around the country began to cast the inept handling of the crisis in racial terms. Terry Ebbert complained bitterly about the lack of federal support, saying they should have arrived within twenty-four hours. Al Sharpton told Tucker Carlson on Fox, "If this had been Palm Beach, the Eighty-second Airborne would have been there Monday afternoon."

Let's face it. Hearing that remark, many if not most Americans immediately thought, "If this had been Palm Beach, there would have been no need for the Eighty-second Airborne because there would not have been any looters." Given the per capita income of such a community, I should hope not, but many Americans would be judging in terms of race, not wealth. The subject is a touchy one with me, as noted in the Introduction. Natal, where I grew up, was and is the most English (as opposed to Afrikaner), and therefore liberal (as opposed to reactionary), province in South Africa, but that didn't stop almost everyone in Natal from having black servants. We had one maid, Stephanie, a Xhosa, and from time to time my father would hire gardeners and day laborers. When I was thirteen or fourteen he bought an old house and rebuilt it over a number of years, hiring some Zulu fellows to help. After school, we, or sometimes just I, would check up on the work and give instructions. So a fact of life I can't deny is that under the apartheid system, and to some extent anywhere in Africa, where most white people have black servants, a white kid grows up instructing black adults how he (or she) wants everything done. I can sometimes be a little cocky, I'm told, and to the extent to which the allegation is true, I consider it a vestige of apartheid. At an early age I learned how to give orders, and it became kind of second nature.

At the same time I was completely aware of what was going on in my strange society. I knew the rest of the world was not organized in the South African fashion. I remember one day at the bus stop, waiting to go home from school. On the bench, certain seats were for whites, others for blacks. The same segregation held for the buses themselves. After this bad day at school, some black kids didn't get up to make room for me on the bench. A spoiled brat, I told them they had to move, and I saw a police car and waved at it. Those kids left. I wish I could find them and apologize. That was a turning point for me. I knew what I had done. I understood the injustice—and it was about time, because my parents, especially my mother, had always worked hard against apartheid. I didn't tell them what had happened.

As a student in Pietermaritzburg in the mid 1970s, a group of us tried to wage the good fight, but the combination of the security police and the government's equivalent of the Patriot Act was very effective in scaring us from acting too boldly. We were harassed at antiapartheid demonstrations. My sister, Gwyneth, who ran adult education programs for rural blacks in Johannesburg, had her office firebombed. But here in Louisiana it was different, wasn't it? I was in the land of the free, so I could do and say something. In New Orleans following Katrina I could hear the desperation in the people's voices—who could fail to?—and the preoccupation in some quarters on the crime question really angered me. The issue was totally overblown and hurt the reputation of New Orleans and Louisiana. There may have been a political agenda behind it as well, because it certainly serves to undercut public support for the relief effort.

Walter Maestri, the emergency manager of Jefferson Parish and one of the best such managers in the country, said many times to many reporters, "They told us they would be there in forty-eight hours. We just have to hang on for forty-eight hours." By my calendar, that meant that time was up on Wednesday. It was definitely up on Friday. Where the hell was the federal government? They were now in charge. Instead of effective action we were confronted by these Haiti-like scenes playing out in the United States. I saw the residue of latent racism. Or not so latent. Fear of looters and thugs, and recognition that their own scarce resources were spread pretty thin, seems to be the reason a crowd of several thousand people, including many elderly and children, were stopped by the police in Gretna, right across the Mississippi River from downtown New Orleans? I picture a different welcome if they had all been white. After those Gretna police officers were sued for that action, the city council voted unanimously to offer them the services of the in-house counsel. Councilman Chris Roberts sponsored a commendation and said that these officers had saved the area from being stormed by looters. He also said that a couple of thousand evacuees had been allowed to camp out until they could be bused to distant points. In a subsequent interview on national television, one of the Gretna officers explained that they had very scant resources of their own—water was short all around—and feared their supplies had run out already.

The worst of these provocations—for me, at least—may have been that *Army Times* piece about "taking the city back" and defeating the "insurgency." This really set me off. It was the ultimate insult to these people struggling to survive. There were some criminals in New Orleans; there are criminals everywhere. What are we teaching our young troops in today's army? Do they play too many violent computer games? Why say and print this crap?

How about Congressman Richard Baker from Louisiana, who was quoted in *The Wall Street Journal* as telling lobbyists, "We finally cleaned up public housing in New Orleans. We couldn't do it, but God did." Subsequently he stated he had been misquoted. But when I read such statements—there would be many equivalent ones to come—and I remembered that this man, specifically, professes to be a practicing Christian, but he obviously does not read the same Bible I do, definitely not the same passage in the Book of Matthew:

> *Then the King will say to those on his right, "Come, you who are blessed by my Father; take your inheritance, the kingdom prepared for you since the creation of the world. For I was hungry and you gave me something to eat, I was thirsty and you gave me something to drink, I was a stranger and you invited me in, I needed clothes and you clothed me, I was sick and you looked after me, I was in prison and you came to visit me." Then the righteous will answer him, "Lord, when did we see you hungry and feed you, or thirsty and give you something to drink? When did we see you a stranger and invite you in, or needing clothes and clothe you? When did we see you sick or in prison and go and visit you?"*

The King will reply, "I tell you the truth, whatever you did for one of the least of these brothers of mine, you did for me." (25:34–40)

I would like to point out that when the cavalry finally made it to New Orleans over the weekend, Lieutenant General Russell Honore issued a "guns down" order. "This is not Iraq," he said. "You are part of a humanitarian relief convoy."

I have a great deal of admiration for this man.

On Tuesday and Wednesday Mayor Nagin predicted that between 2,000 and 10,000 had died or would soon die in the city. He was criticized for this remark later, but the estimate was not a ridiculously wild guess. Before the storm many reporters had repeated the estimates from the American Red Cross that between 25,000 and 100,000 people might die because they could not or would not evacuate. Plus, Nagin was using his bully pulpit to attempt to speed the federal and state response to his flooded city. I can't hold that against him, and as Mark Schleifstein pointed out in the *Times-Picayune*, "Just think what the national media would be saying today if Nagin had said only 100 were feared dead—and it turned out there were more than 1,000." I think the mayor was also listening to some of our people at the hurricane center who, in the frenzy following the levee breaches, were attempting to estimate how many people may have been left behind by "backing out" the number from a variety of statistics. An estimate was needed in part to help guide the first responders who were combing New Orleans and other areas for survivors. (Our team was also able to plug all 911 calls into a GIS database that translated them into dots on maps. We could then correlate that information with Ahmet's flood-depth map, to help direct the rescuers.)

The fatality numbers I kept hearing behind the scenes—hundreds of deaths—just didn't add up for me. I've cited the public opinion surveys showing that about 70 percent of the 1.3 million residents of metropolitan New Orleans would evacuate under the threat of a storm such as Katrina. The citizens actually beat that figure, apparently, with greater than 75 percent fleeing the area, as I've reported, but I didn't know this. Therefore we were calculating close to 400,000 people staying behind. If there were, at the very most, 50,000 people at the Convention Center and the Superdome combined—and there were probably nowhere near that number; more like 30,000—where were the remaining 350,000 people? Or assume that the city had achieved a world-record 90 percent evacuation rate, which might have been the case, given the numbers of cars counted as they left town, and assume the highest possible number of refugees at the two makeshift shelters, there were still perhaps 80,000 people unaccounted for. Some of these would have been in the parts of Jefferson Parish that had not been flooded. Let's say they numbered 50,000. We still end up with at least 30,000 missing. With this analysis, our graduate student Ezra Boyd, who was developing a "flood fatalities model" funded by us for his Ph.D., thought the estimates of hundreds of deaths must surely be very low.

Ezra's model is the number of fatalities divided by the exposed population, and his hypothesis assumed a correlation between fatalities and water depth. His data from various floods around the world reveals this to be the case. Given the high water depths for Katrina, Ezra predicted a one-third fatality rate. With an estimated 30,000 people exposed to the flood—the lowest plausible number, pulling the calculation from the preceding paragraph—we could expect about 10,000 deaths. A more comprehensive but similar calculation that accounted for the range of estimated flood depths throughout the region lead to an estimate of perhaps 20,000 potential fatalities. That was the number initially provided to the state and local response agencies, and it was much too high. The *lower* number was even too high. Why?

Ezra's calculations did not account for two crucial factors: An evacuation far more successful than anyone had imagined and the effectiveness of the search and rescue operations. Indeed, it is clear that local first responders achieved remarkable success following

Katrina. The effort was not textbook, it was anything but coordinated, but it saved many, many lives. If it was not for these thousands of heroes, the number of fatalities could have been 10,000.

Six months later I'm still surprised at the quasi-official final figure of 1,300 deaths along the entire coastline, 1,100 in the New Orleans area. I hope it's right, but this number may be too low. The final answer may be considerably higher. Over 3,000 persons are still reported missing; of these almost 500 are believed to have drowned, based on a home address that was either totally destroyed or completely flooded. As of February 2006, we believe the upper limit for fatalities is 3,000. Let's hope we don't get there.

What about all the other numbers bombarding us that week, most of them enormous? While looking at the pictures of desperate and stranded people we heard about 13.4 million liters of water at the city limits, apparently, along with 10,000 tarps, 5.4 *million* ready-to-eat meals, 3.4 *million* pounds of ice, 135,000 blankets, and 144 generators. Tons of everything were ready to go. The Coast Guard, the National Guard, the Air National Guard, the Louisiana National Guard, the state police, the Army, Navy, Air Force, Marines—everyone was coming, but with the exception of the hundreds of rescue boats in the waters and a few helicopters in the air, there did seem to be an awful discrepancy between the resources in personnel and supplies allegedly available and those actually on hand and helping people. On Thursday night, just as I was suggesting to Larry King that the wheel nuts had come off, Mayor Nagin lost his cool on WWL radio and said about President Bush, "We [have] an incredible crisis here and his flying over in Air force One does not do it justice. Don't tell me forty thousand people are coming here. *They're not here.* It's too doggone late. Now get off your asses and let's do something and let's fix the biggest goddamn crisis in the history of this country. . . . Excuse my French, everybody in America, but I am pissed." Nagin later made clear that he meant no disrespect to the president, he was just at his wit's end. This was his "desperate SOS."

On Friday we did see signs of progress. The airport was functional again. A military convoy—the cavalry that the mayor had always been depending on—rolled into town with major fanfare from the networks just about as another convoy rolled out of town—buses with the folks who had been stranded at the Superdome who were now on the way to the Astrodome in Houston. That nightmare was coming to an end for twenty thousand or more people, and by late Saturday evening the building was nearly empty.

What about the Convention Center three miles away, which was never an official shelter but ended up with perhaps fifteen thousand people by the end of the week? This was one of the biggest puzzles of the whole first week. The *Times-Picayune*'s invaluable Nola.com Web site may have been first outlet to reveal that several thousand folks were stranded there, presumably refugees from the nearby central city neighborhoods. This was on Wednesday morning. On Thursday—over twenty-four hours later—Homeland Security chief Michael Chertoff said on NPR's *All Things Considered*, "I have not heard a report of thousands of people in the Convention Center who don't have food and water." That same night—thirty-six hours after the *Times-Picayune* revelation—FEMA chief Michael Brown told Paula Zahn on CNN that he had heard about these people only a few hours earlier. Zahn was incredulous, and said so. From that moment, "Brownie," as President Bush called him, the man who had been the stewards and judges commissioner of International Arabian Horse Association for ten years before joining FEMA as general counsel in 2001, was probably doomed. On *Frontline* two months later Brownie claimed to have misspoken in that interview with Zahn. When reporter Martin Smith noted that he had made essentially the same remark on three separate occasions, Brownie bristled and challenged Martin Smith to switch chairs with him and see how he liked it. No matter. On the Thursday after Katrina struck, his remark to Zahn fixed in a lot of minds, including my own, the image of an agency, or at the very least a leadership, incomprehensibly and inexcusably out of the loop.

On Friday we read about the corpse in the wheelchair outside the Convention Center, with the woman's name hand-printed on a card:

Ethel Freeman. The Reverend Isaac Clark said, "We are out here like pure animals. We have nothing. . . . Billions for Iraq, zilch for New Orleans." That day the first significant supplies of food and water finally reached these people. Some aid agencies said they had waited for three days, their vehicles packed with water and food, because FEMA had said not to go in unless they had the necessary paperwork from the state. The state knew nothing about this requirement, nor was there any such paperwork—and even if there had been, go anyway, for God's sake!

On Saturday the first buses rolled up to the Convention Center and the last of the big evacuations from New Orleans finally got under way.

SEVEN

IS ANYONE IN CHARGE HERE?

If I had been completely cut off from all news during that first week except for what I picked up from the media's questions, I could have followed the general course of events and had a pretty good idea of the evolving public perceptions and reactions. Over the prestorm weekend most of the questions had been of the "How bad will it be?" variety. But as early as Wednesday—two days after landfall—I was asked more and more about the warnings that I and many others had been delivering for years. I was asked about FEMA's Hurricane Pam exercise the preceding year, which had been designed to prepare everyone for a storm much worse than Katrina. With the flyover by the president on Wednesday and the assorted astounding remarks by FEMA's Michael Brown in the days to follow, I could tell where this whole story was headed: Where was the government? Where was FEMA? Where was the *plan* to handle this long-predicted emergency?

Part of the answer is conveyed by a brief history of FEMA—officially, the Federal Emergency Management Agency—which was created by President Carter in 1979 under pressure from the state governors, who had wanted better federal help with disasters such as Hurricanes Betsy in New Orleans in 1965 and Camille in Mississippi in 1969, the earthquake in northern California in 1971, and Hurricane Agnes in the mid-Atlantic states and New England in 1972. Carter's executive order folded into the new agency some of the one-hundred-odd agencies that were nominally involved with disaster relief. Everything I've read holds to the line that upper management positions at FEMA quickly gained the reputation of being a suitable reward for midlevel political hacks, sort of like an ambassadorship to some cold capital in central Asia, although this is unfair to the first chief, John Macy, former director of the Civil Service Commission, former president of the Corporation for Public Broadcasting, and then former president of the Council of Better Business Bureaus. Macy himself was not a hack by any means, but neither was he a professional disaster expert. Neither were Ronald Reagan's two appointed directors or George H. W. Bush's one.

In 1992, when Hurricane Andrew destroyed or severely damaged 125,000 homes in south Florida, FEMA was prepared to do essentially nothing, and that's what it did for days. Local outrage eventually stirred the first Bush White House into action, but it was too late to staunch the criticism, and the affair didn't help his reelection campaign against Bill Clinton. He carried Florida, but not by much, and the FEMA fiasco was partially blamed. (A quarter of a million people ended up living in FEMA's mobile homes in Florida for two years or longer.) Clinton won that presidential election and quickly took a tip from the big-city mayors in the Northeast, who know that the first thing they have to do to ensure reelection is to get the streets plowed after the snowstorms. Clinton brought in James Lee Witt as the new FEMA chief, the emergency response chief from Arkansas (and also his friend, to be sure). Witt was the first chief with extensive experience in the field. His top deputies had run regional FEMA offices. They were professionals. (Brownie's number-two man during Katrina was a promoted "advance man" for the 2000 presidential election. Next in line was a PR man for that campaign.) Clinton also elevated FEMA to cabinet-level status. Everyone acknowledges that Witt transformed the agency into a professional organization that did a credible job with the flooding of the Mississippi River in the Midwest in 1993, the Northridge earthquake in southern California in 1994, and Hurricane Floyd in North Carolina in 1999. The most innovative work of FEMA in those years was Project Impact, a nationwide mitigation initiative to help change the way this country deals with disasters. Instead of waiting for them to occur, Project Impact communities initiated "mentoring" relationships, private and public partnerships, public outreach and disaster mitigation projects that would preemptively reduce damage. They revised local building and land-use codes, and some even passed bond issues to pay for new measures that would help an entire community. FEMA calculated that the program saved three to five dollars for every dollar spent. From its inception in 1997, nearly 250 communities and 2,500 business partners embraced Project Impact.

In 2001, President George W. Bush appointed Joe Allbaugh to take over FEMA. Not a disaster man; instead, Bush's chief of staff while he was governor of Texas, then his campaign manager in the 2000 presidential election. Among Allbaugh's first moves was dropping Project Impact as an "oversized entitlement program." He said he intended to deemphasize the federal role in emergency preparedness, then he bizarrely told the *Times-Picayune*, in an interview for its 2002 award-winning Washing Away series, "Catastrophic disasters are best defined in that they totally outstrip local and state resources, which is why the federal government needs to play a role."

Read it and weep. The contradictions in those attitudes reveals a mentality that completely fails to understand that poor preparation for a predicted disaster exacerbates the problems of coping with the aftermath of that disaster exponentially. Yet how could anyone in authority in the disaster business miss this necessary correlation?

In the weeks following the September 11 terrorist attacks, FEMA got high marks across the board for coordinating the search-and-rescue operation. Its budget exceeded $6 billion. Then it was folded into the new Department of Homeland Security, where, as a small agency with a smaller budget of only $4.8 billion, it sank almost without a trace. Allbaugh himself told *Frontline* that the FEMA of 9/11 was not the FEMA that failed to handle the Katrina disaster. Even Brownie Brown, who had taken over the reins in 2003, complained that his little outfit had suffered from the "taxes" imposed on it under the new structure. Two months before Katrina he had circulated a memo complaining about the ever lower status of his agency and the lack of funding for any preparedness function whatsoever. (In October, after Katrina, Bush administered the coup de grace in this respect, removing any preparedness role from FEMA and giving it to the Preparedness Directive in the Department of Homeland Security. FEMA is now strictly a response organization. Again, are these people thinking about the relationship between preparedness before the fact and response after the fact?)

Worse for FEMA, terrorism became the main focus for the department. Homeland Security has handed out over $8 billion to the states to buy just about anything they want in the name of antiterrorism. We all know the stories: bulletproof vests for the fire department dogs in Columbus, air-conditioned garbage trucks in Newark, self-improvement seminars for sanitation workers in the District of Columbia. States and localities far removed from any likely terrorist attack have received a much higher percentage of the funds than makes any sense whatsoever.

We call this pork, which is really just a polite word for corruption. Call it "soft" corruption.

What was FEMA's specific plan for rescuing New Orleans from the disaster that the agency itself had predicted for years? One day a group of academics or reporters, or both, is going to take it on themselves to try to penetrate the bureaucratic barriers and dissect it in painstaking detail. As I've said, that's going to be quite a book and quite a challenge to put together. The job will take years. For one thing, there is no there there: FEMA is really a coordinating agency. Prior to being moved into the Department of Homeland Security it tried to bring together the activities of at least sixteen different major programs in different departments of the federal government, plus all the state and local programs. Maybe the new organizational chart makes the coordination under DHS easier on paper, but the experience on the Gulf Coast in 2005 makes me wonder. Presumably, the agency itself will conduct such an investigation of its performance, but no bureaucracy investigating itself is likely to provide the answers we need.

Of course, Homeland Security officials defend FEMA's readiness, pointing, for example, to the five logistics centers around the country. Formerly, there were sixty-five. Is five necessarily better than sixty-five? Maybe, maybe not. By way of excuse for the absence of supplies on the ground immediately after Katrina struck, we heard time and again that FEMA couldn't preposition its forces directly in the path of a major hurricane. True, of course. So where were they prepositioned? We know that Wal-Mart activated its emergency command center in Bentonville, Arkansas, on Wednesday, August 24, five days before landfall. It was staffed the following day, and emergency supplies were shipping out on Sunday—the Sunday before landfall. Where, *exactly*, were all of FEMA's rescue teams, supply teams, supply caches, and all the rest on the Sunday before the storm? How about on the Thursday *after* the storm? What auxiliary communications systems were ready to go? (Brownie told *Frontline* that FEMA did have communications, "Just not enough.") What plans were in place to get the New Orleans airport up and running immediately? (What was one of the first facilities secured in Iraq? The Baghdad airport.) Where were the transport planes packed and waiting at airports out of harm's way. What facilities had been preselected for handling a million evacuees for a long period of time? (I know that answer. None.) Forget the published organizational charts, the flowcharts, the operational procedures documents. Where were the actual supplies and people, the wings and the wheels? How was everyone going to communicate? What were people *doing*?

I want to emphasize that I don't for a moment underestimate the enormity of the challenge FEMA faced with Hurricane Katrina. This was an immensely more complex job than the one the agency had faced in New York City following the 9/11 attacks, where the disaster area was tightly defined and immediately accessible, where every service was intact just two blocks away, with no vast diaspora of refugees. There's no comparison, really. More than two million evacuated during Katrina. One million were displaced from their homes for months, at minimum. Half a million were in shelters in the first two weeks of the disaster.

In fact, it would be easy to look at such numbers and think, well, no wonder FEMA and every other agency were overwhelmed. But that's an incomplete analysis, because this was not a catastrophe out of the blue—a tsunami, say. A hurricane like Katrina and the consequences were a given. The big numbers were not a big surprise, or any surprise at all. Plan for them, just as an army plans for a war (presumably), that is, with mind-numbing detail. Look at our huge effort in staging for the invasion of Iraq. Every contingency was thought through. (I'm talking about just the war here, not the ensuing occupation, many aspects of which might have been prepped by FEMA itself. In fact, it would be pretty easy to draw an analogy between the government's failed preparations for the predicted disaster of Katrina and the botched occupation of Iraq. War we're good at. The best. We stand alone. But then what? Of course, questions were raised in Louisiana about the fact that roughly 40 percent of the state's seven thousand National Guards were on duty in Iraq. No effect at all, we were told. Nonsense. Every guardsman I spoke to bemoaned the fact that their units were shorthanded and without enough people to do security, rescue, and logistics.)

Which brings us back to the Hurricane Pam exercise in July 2004, an intensive one-week war game, in effect, put together by a Baton Rouge consulting company on behalf of FEMA for the sole purpose of getting ready for the big one. Just the month before, William Carwile, FEMA's coordinating officer in Mississippi, had written Michael Brown an eleven-page memo stating that the agency's teams of national response managers were unprepared and receiving "zero funding for training, exercise, or team equipment . . . [and these responders] provide the only practical, expeditious option for the director to field a cohesive team of his best people to handle the next big one." AP reporters turned up this document in December 2005. It is full of other highly critical assessments. But maybe the Pam exercise would provide a way forward for FEMA. Marc Levitan was asked to sit on the committee setting up the exercise, and we were excited. FEMA itself—the agency that would be in charge of the disaster waiting to happen in New Orleans—had realized the necessity of planning and coordinating the work of all the other agencies involved. Apparently it understood the requirement of assessing the readiness of those agencies. Moreover, this was an opportunity to get our science out on the table. We had been officially studying the New Orleans peril for three years. We had looked at the hospitals' jeopardy and the problem of the shelters; we had conducted public opinion surveys; we had looked in great detail at the public health issues and had met more than once with the people at the Centers for Disease Control. All in all, we had an inventory, so to speak, of what would be at stake. Quite honestly, no one anywhere knows anything close to what we know about this subject. Nor could they be expected to. This has been our main job.

Naively, perhaps, we were primed to be real partners in the event. Marc had great hopes that we would be fully engaged in the exercise. But that was not to be. He is a very persuasive guy, very diplomatic, but as hard as he tried he could not convince the consulting firm FEMA had hired to broker the exercise that we had already developed with our GIS database much of what they needed. No, they replied, they would produce "all the necessary maps and such data." Our role was confined to producing a storm-surge animation for the fictional Hurricane Pam, a slow-moving Cat 3 that passed just west of New Orleans, flooded the city, and left sixty thousand dead. (The actual track was developed by the National Weather Service, with Marc designated as the LSU person liaising with them. The National Weather Service used its SLOSH surge model to get the flood just right; when they were satisfied, we did our ADCIRC animation.)

Colonel Mike Brown (no relationship), who was the assistant director of the state's Department of Homeland Security at the time, and who I'm sure was one of the prime movers in getting FEMA to stage this exercise, had always been one of the LSU Hurricane Center's strongest supporters. He had real-world experience that gave him an appreciation of the value of our research. At his bidding, our Hurricane Public Health Center was tasked with providing part of the introductory briefings for the exercise, which involved a couple of hundred people altogether from many different local, state, and federal agencies. Paul Kemp gave one briefing, I the other. We itemized everything we knew about the peril, right down to the fire ant issue. For the weeklong exercise itself I spent most of my time with the search-and-rescue planning team, made up mostly of state Department of Wildlife and Fisheries and Coast Guard officials. I felt at the time that these folks had a grip on the issues and got a lot done, and I think their performance during Katrina proves the point. As soon as the winds dropped Monday afternoon, they were out there—Wildlife and Fisheries, Coast Guard, Louisiana National Guard, and the various fire and police departments. Boats and supplies had been effectively prepositioned, including at the Jackson Barracks in Arabi, just outside the Lower Ninth, as I've described.

The search-and-rescue exercisers discussed how to deal with firearm-toting pirates who might commandeer a rescue boat to ensure their families got out of the flooded city first. Since this was Louisiana, where boat ownership is extremely high, they envisioned the large fleet of potential freelance rescuers (and some looters as well) who would probably stream out from the north shore of Lake Pontchartrain in something of an Operation Dunkirk. In the real one, British citizens had used their own craft to cross the twenty-three miles of the English Channel to rescue their soldiers trapped by the Germans on the French shore at Dunkirk—this at the beginning of World War II. For Hurricane Pam the exercisers set up predetermined pickup sites on the south shore of the lake—the city of New Orleans. Almost every boat in the state is equipped with a marine VHF radio, so we discussed the possibility of the Coast Guard's erecting a portable VHF tower if necessary. In the event, I don't know if this was necessary during Katrina, but certainly when I crossed the lake five days after landfall, the VHF communication system was alive and well. It turned out that the Katrina disaster, severe as it was, paled in comparison with what a storm like Hurricane Pam would do to New Orleans, so a full-fledged "Operation Dunkirk" was not set up. (Aaron Broussard, president of Jefferson Parish, did invoke the memory of the famous operation when he called for citizens with flat boats to help in the rescue.)

At one of the first meetings on the first day of the Pam exercise, I felt—*we* felt; the LSU group—that some of the maps weren't quite right. There were also some problems with the flood depths, as I recall. Anyway, I asked one of our mapmakers if we could generate a better map on short notice. It wasn't six hours before I got a call from someone with the state telling me to back off. The consulting company was very upset; allegedly, we were jeopardizing the whole exercise. So we backed off. Very frustrating, since we had all the information. But after a few days we ended up sharing a lot of our GIS data anyway. We produced flooding maps, we made posters. We were good scouts. We would see a need, rush back to LSU and generate the required map, and then bring it back to the exercise.

A day or two into the exercise, Colonel Brown asked me to put together a CD with the storm-surge models, examples of the GIS data, and background information. This request really helped our overall cause, I think, and we produced the disk overnight and handed out copies right and left. Then the colonel asked me to brief a White House representative (from the Office of Management and Budget, I think) who had just arrived. I outlined everything for him—evacuation, housing, contamination—but I don't think it had much impact. I got the sense that several federal officials were in attendance only because that's what was required. Perhaps not everyone at the exercise believed this catastrophe could actually happen. Perhaps the ideologues didn't think it was a federal responsibility. But such explanations don't account for the attitude of the FEMA people, who knew very well that this could happen and who, presumably, did understand it would and should be a federal responsibility. On numerous occasions, when we tried to interject some of our science, we were bypassed or even ignored by these knowledgeable people. As I said in the introduction, I just don't think some of them took us seriously. I, specifically, was the foreign geek who talked about tent cities or some such *organized*, short-term housing for the evacuees, and this was simply not in the cards—one of several critical blunders during the Katrina disaster response.

The success of the rescue efforts conducted by Wildlife and Fisheries, the Coast Guard, and the other first responders reflected the benefits of the Hurricane Pam exercise. I think it convinced *almost* everyone involved of the enormity of the challenge posed by a flooded New Orleans. However, in many other emergency response aspects, the exercise fell short. We know this for the simple reason that we know what happened—and didn't happen—during Katrina. During the Pam exercise there was discussion of the problem of evacuating the 127,000 people in New Orleans without access to vehicles. "To be determined at a later date" was the solution reached during the exercise, and I can only conclude that this same solution pertained to numerous other issues, because they never got determined at a later date, this despite the express intention to keep the "fairy dust" to a minimum. In fact, as I mentioned to Tim Russert, FEMA representatives had talked about a second Pam exercise to focus on the low-mobility groups. It didn't happen.

Now I want to amplify on the point I made while discussing the nonevacuation of the low-mobility groups over the weekend before Katrina: Once President Bush declared a national emergency on Saturday, August 27 (retroactive to the day before), the federal government was in charge, pursuant to the Stafford Act of October 2000. That act states (§401 and §501): "All requests for a declaration by the President that a major disaster or emergency exists shall be made by the Governor of the affected State. . . . As part of the request, the Governor must note that the State's emergency plan has been implemented and the situation is of such severity and magnitude that the response is beyond State and local capability and Stafford Act assistance is necessary."

Governor Blanco complied with this requirement on the Saturday before the storm. (Michael Brown would actually tell a House investigating committee in November 2005 that New Orleans was explicitly exempted from that request. Why would he say that? I'm looking at the official requests. There's no exemption for New Orleans.)

Once the governor's request is made through the regional FEMA office, state, local, and federal officials are enjoined to estimate the extent of the disaster and its impact on individuals and public facilities. This had definitely been done. I need only quote Michael Brown himself, who told CNN just days after landfall, "We actually started preparing for this two years ago. We had decided to start doing catastrophic-disaster planning, and the first place we picked to do that kind of planning was New Orleans, because we knew from experience, based back on the forties and even the late 1800s, if a Category 5 were to strike New Orleans just right, the flooding would be devastating. It could be catastrophic. So we did this planning two years ago. And, actually, there's a tabletop exercise with the Louisiana officials about a year ago. So the planning's been in place now. We're ready for the storm."

If the Stafford Act wasn't sufficient to establish federal hegemony—which it was—on Tuesday, the Department of Homeland Security declared Katrina an "incident of national significance," triggering the first use of the official National Response Plan put in place after September 11 and designed to bring all military and civilian units together in a seamless operation to confront a catastrophic event, be it natural or man-made or both (as in Katrina). In the earlier discussion of this subject I mentioned Ira Glass's interview with author William Nicholson on NPR's *This American Life*. I quote that interview again:

> NICHOLSON: Well, basically the way it works is, the Secretary of Homeland Security designates this as a catastrophic incident, and federal resources deploy to preset federal locations or staging areas, so they don't even have to have a local or state declaration to move forward with this.
>
> GLASS: And in other words, it doesn't matter what the governor says, it doesn't matter what the local people say, basically, once that happens, they can just go ahead and do what needs to be done to fix the problem.
>
> NICHOLSON: That's correct. It's utterly clear that they had the authority to preposition assets and to significantly accelerate the federal response.
>
> GLASS: And they didn't need to wait for the state?
>
> NICHOLSON: They did not need to wait for the state.

Secretary Michael Chertoff was now officially in charge of the largest domestic relief effort in this nation's history. As someone said, "It's exam time." Yet Chertoff would claim that America's "constitutional system" gave primary authority to each state. And Brownie would declare repeatedly that the states were always in control, that he didn't have the resources to take over, that FEMA's attitude was "'tell us what you need.'" In the remarkable *Frontline* interview he told correspondent Martin Smith that all along he had wanted to put the blame where it belonged, on Mayor Nagin and Governor Blanco, but he thought that wouldn't be appropriate or helpful. Brown was utterly unrepentant in that interview, although he did state that his worst mistake was not getting massive troop support instantly. In January 2006 he had become a tiny bit more willing to take responsibility. Speaking to a convocation of meteorologists, he acknowledged his failure to communicate the enormity of the disaster and to request assistance from the military sooner.

Hearing all this, I thought of an editorial in the *Biloxi Sun-Herald*, which I read about in a Paul Krugman Op-ed column in the *New York Times* titled 'CAN'T DO' AMERICA. The *Sun-Herald* editorial tells us that reporters were listening to the terrible stories of evacuees at a local junior high school shelter while looking across the street at Air Force personnel engaged in basketball and calisthenics. How dispiriting is that? As Knight-Ridder reporters revealed in an excellent article dated September 11, a 1993 report from the Government Accountability Office, the investigative arm of Congress, had concluded that the Department of Defense "is the only organization capable of providing, transporting and distributing sufficient quantities of items needed" in a major catastrophe. Yet the Pentagon remained aloof during this one in Louisiana. Secretary of Defense Donald Rumsfeld attended a baseball game in San Diego on Monday night. The boss of bosses was on vacation in Texas. No one in the Defense Department set up a Katrina task force until Wednesday.

In December 2005, however, FEMA officials still stuck to their guns, telling the Senate investigating committee that, yes, they had been overwhelmed, but local and state officials had failed to do their jobs with evacuations and shelters. At the Superdome state police had turned away two FEMA trucks loaded with food, water, and ice. That's what the FEMA folks told the senators, but I find any such story absolutely incredible. I would have to see the proof. What conceivable motivation could any state trooper have for such an action? FEMA wanted the state of Louisiana to tell it what was needed? Fine. The state forwarded a list forty-eight pages long—but the senators were told that FEMA got "hundreds and hundreds of requests in addition to valid requests. This is just a lack of understanding of what FEMA does."

Hold everything. If there was such a fundamental misunderstanding, whose fault was that? What agency is the overall coordinating agency in a national disaster? What federal agency was running the Pam exercise? Wasn't the major purpose of that exercise to understand exactly who would be doing what during an actual hurricane emergency? What agency's job is it to know and to plan for contingencies?

Michael Chertoff, the head of Homeland Security and Brownie's boss, referred to Katrina as an "ultra catastrophe" and a "combination of catastrophes that exceeded the foresight of the planners, and maybe anybody's foresight." This was two disasters in one, he said—the storm itself, then the flooding—and he blamed the flooding for the logistical problems. FEMA's Natalie Rule said the same thing: "It is not as simple as driving right up into the city of New Orleans and starting rescue as we might be able to do in other disasters, such as an earthquake."

Dissimulation or ignorance? Hard to say, but to those of us watching and listening, such statements were almost too much. On the Wednesday after the storm Brownie was honest enough to tell Larry King that the storm "caused the same kind of damage that we anticipated." Of course it did! The Hurricane Pam exercise theoretically planned for a catastrophe much worse! Brownie had returned from a tour of the tsunami damage in Asia and informed a major meeting that the most likely roughly equivalent catastrophe in the United States was . . . New Orleans. His agency had anticipated catastrophic flooding, staffed for catastrophic flooding, prepped for catastrophic flooding, war-gamed for catastrophic flooding—only to be completely overwhelmed by the real thing.

The *New York Times* reported that the federal government had foreseen the flood risk but not the levee failure. Fire whoever said this to the *Times*, because the distinction is absolutely ridiculous. As far as FEMA's planning was—or should have been—concerned, whether the standing water in New Orleans following any hurricane came from overtopping of the levees or failure of levees, the resulting catastrophe would present *exactly the same problems.* The Army Corps of Engineers would face a very different set of issues, but not FEMA. President Bush added to the confusion or obfuscation on Thursday when he said to Diane Sawyer on ABC's *Good Morning America*, "I don't think anybody anticipated the breach of the levees. They did anticipate a serious storm." A White House spokesperson explained later that the president meant that no one anticipated breaches *after* the storm had passed. In the first place, the breaches occurred just as, or even slightly before, the storm passed to the east of the city. In the second place, again, this distinction is utterly irrelevant. We at LSU had predicted the flooding; FEMA had brainstormed and war-gamed the flooding for years. We know for a fact that the news had penetrated the walls of the White House, thanks to Joby Warrick's reporting in the *Washington Post* in late January 2006. Forty-eight hours before Katrina struck, FEMA had laid out for White house officials a computer slide presentation comparing this storm's likely impact to that of Hurricane Pam—except that Katrina could be worse. At 1:47 A.M. on Monday, August 29, five hours before the storm hit, Homeland Security's National Infrastructure Simulation and Analysis Center e-mailed a forty-one-page assessment to the White House "situation room," repeating the dire warnings of breached levees, massive flooding, and major losses of life and property.

Tim Russert would challenge Secretary Chertoff on *Meet the Press* about how the president "could be so wrong, be so misinformed." The answer was not illuminating.

Then, on March 1, the AP published White House documents, including a video, showing that President Bush was in fact *not* so misinformed after all. In a briefing the day before Katrina hit, he had been clearly apprised of the imminent and possibly catastrophic danger in New Orleans. He didn't ask a single question, true, but he was told what was about to happen. The other major revelation from the documents is Michael Brown's apparent engagement with the crisis. He appears to be blunt, anxious, visibly concerned.

Senator Mary Landrieu, whose father, a former mayor of New Orleans, had evacuated the city for the first time, said at the Emergency Operation Center that she had difficulty convincing FEMA what just one breach in the levees could mean. She said Michael Brown "had a difficult time understanding the enormity of the task before us." Yet the video unearthed by AP shows a clearly focused Brown. He had told King that the damage was anticipated, as quoted above, and he had warned months earlier that the French Quarter could end up eighteen feet underwater (which it didn't).

It was all so bizarre and, yes, maddening. Then over the following weeks came the reports about Brownie's incredible e-mail exchanges throughout the period. On the Monday of the storm he wrote, "Can I quit now? Can I go home?" A couple of days later, "I'm trapped now, please rescue me." Should we interpret such remarks as facetious quips meant to relieve the pressure, as clueless jokes, or as unbelievably callous disregard? I don't know. Thirty minutes after he wrote an e-mail saying that no action had been taken on the use of airlines to evacuate people, his deputy director, Michael Lowder, wrote, "This is flat wrong. We have been flying planes all afternoon and evening." When FEMA's Marty Bahamonde wrote Brown that the situation was "past critical," his boss replied, "Thanks for the update. Anything specific I need to do or tweak?" Then there were the famous e-mails to and from one of his PR staffers concerning what clothes he should wear for the TV interviews. He was advised to roll up his sleeves—literally—and look busy.

But enough. Making sport of these people and their actions and inaction is too easy, and nothing about the Katrina catastrophe should be a laughing matter. People were dying.

It's possible for an organization with dazed and confused chiefs to nevertheless function well enough down the ladder, but this did not seem to be the case this time with this agency. According to Aaron Broussard, president of Jefferson Parish, Wal-Mart had tried to deliver three trailer truck loads of water to his people, only to be turned back by FEMA. Walter Maestri, the parish's superb director of emergency preparedness, said bluntly that FEMA reneged on its offer to relieve county emergency staffers forty-eight hours after landfall. In conference calls with FEMA on the Saturday before the storm, Jefferson Parish officials, including Maestri, went over detailed lists of equipment and manpower they would require immediately. Don't worry, the cavalry will be there within forty-eight hours: That was still Maestri's belief. According to Coast Guard officials, FEMA stopped the same parish from acquiring one thousand gallons of diesel at the Coast Guard station at Belle Chase, just down the road. FEMA's William Carwile wrote in an e-mail from Mississippi: "Biggest issue: resources are far exceeded by requirements. Getting less than 25 percent of what we have been requesting from HQ daily." On Friday, Bill Lackey, a FEMA coordinating officer in the disaster area, admitted bluntly, "It seems our planning was inadequate. We worked on it, we exercised for it, but the reality of it—we've been working as hard as we can."

The newspapers and the airwaves were crowded with such stories. And why did the TV crews seem to have such good access to most areas of New Orleans while FEMA didn't? The rest of us shook our collective head in wonder and dismay and anger. President Bush said to his FEMA chief, while touring Mississippi and the airport in New Orleans, "Brownie, you're doing a great job." (In Mississippi, he also lamented the loss of Trent Lott's shorefront estate and reminisced about all the great parties he had enjoyed on Bourbon Street, back in the day, when that's what he did.) The president's praise for Brownie on Friday morning followed by less than a day Brownie's admission to Larry King that he'd just learned about the thousands of people stranded at the Convention Center. Shortly after bestowing these kudos, the president tacked and said that the results of the federal effort were "not acceptable." That afternoon, on his way to the breach at the 17th Street Canal, he asked local emergency managers if they knew about the Hurricane Pam exercise. Of course they did.

"How did you use it?" the president asked Walter Maestri of Jefferson Parish.

"We used it as a bible."

"I'm sorry. Heads will roll."

EIGHT

WETLANDS FOREVER

Michael Brown, FEMA, evacuations, levees, floods, drownings, and death—we wouldn't be talking about any of this—I wouldn't be writing this book—the tragedy of Hurricane Katrina would never have happened were it not for the unique, precarious setting of New Orleans and southeastern Louisiana in this small wedge of the North American continent. For once, perhaps, a PR slogan is no exaggeration: Louisiana really is "America's Wetland." The state has 40 percent of the total coastal wetlands in the Lower 48. These four million acres of marshes and swamps and estuaries are one of the world's great ecosystems. For millennia the Mississippi River has supplied the immense resources of freshwater, nutrients, and sediment with which to build this vast expanse, which natural processes of erosion then break down. This dynamic interplay of land and water, where new lands are continuously built and old lands changed and lost, has produced an environment with an unsurpassed diversity in vegetation, wildlife, and fisheries, and an extraordinary biological productivity. On the other hand, it also poses problems for the society that chooses to live here. As the geographer Pierce Lewis put it so well, New Orleans is an "impossible but inevitable city": A port near the mouth of the Mississippi River is a given, but no land near this mouth is suited for the purpose.

The wetlands are absolutely vital for protecting this whole part of the state from any hurricane's storm surge. Along with the barrier islands, they are the best, most natural, least expensive buffer available. We can't understand why the Katrina tragedy happened, or the complexity of the rebuilding problem now facing us, without understanding this domain. It's time for a brief excursion into the local bush, where the real danger is not the alligators but the politics.

Of course, I knew almost nothing about the place when I first arrived at LSU as a graduate student from South Africa in January 1977. I saw the raw, unadulterated wetlands for the first time about a month later, when my supervisor, the wonderful teacher and mentor Harry Roberts, escorted me to the Atchafalaya Delta seventy miles south of Baton Rouge. Wow. I'd never seen such a landscape: bleak but exciting. It was a cold, gray day. The land smelled fresh—and it should have, because it was literally new land—land that was still growing. I immediately loved my new home away from home. For one thing, I've always been fascinated with water, any flowing water. As a boy I'd let a tap run over piles of sand and make little rivers and canals and, yes, levees, though I didn't think of them as such at the time. My mates and I made crude canoes out of corrugated, galvanized tin—the standard roofing material in southern Africa then and possibly now. We'd pound the corrugations and irregularities out of the tin until we had a flat sheet, wrap the metal on a kind of a frame made of wood, scrounge some leftover bitumen tar from construction sites on the roads, heat it up, pour it over the nail holes, and have a canoe. More than one. We put these rough-hewn craft on top of soapboxes made with old push pram (baby stroller) wheels and rolled them to the nearby ponds. When we were through canoeing for the day we'd sink the things, then bring them to the surface the next time, no mean task for small boys. Those canoes were totally unsafe; they had no floatation at all.

In apartheid South Africa in the sixties all young white men were conscripted into the army immediately after graduation from high school. I was luckier than most guys, in that I ended up in an engineering regiment and getting a land surveyor diploma, training that would serve me well. My parents wanted me to be an engineer, for some reason, so I studied engineering for two years at the University of Natal, but there was too much math. I called it quits and went to work at Lake St. Lucia in Zululand. That was in 1969, a time of exploration for me, living in a really wild swamp area in the bush, getting to know the Zulus much better, learning to speak some Zulu. They dubbed me "Ma'thatatini Nge en jeti," which means "someone who drives like a grasshopper," a comment about my method of driving a four-wheel-drive Land Rover along the muddy tracks in the swamps. Those were papyrus swamps, of which there are none in the United States. (By the same token, there are no cypress swamps like those of the southern United States in South Africa.) Lake St. Lucia is one of the last true African wildernesses, an incredibly diverse region, ecologically, geomorphologically, and geologically. You always have to be on the alert for large Nile crocodiles, fellows who see you as a meal. Over five hundred hippopotamus also make the Wetland Park their home, and they are also known to charge the odd boat.

The whole wetland complex was also an area undergoing enormous stress due to bad management practices in some of the river valleys feeding the system. Here I found my passion, one of them: understanding how natural systems function and how to correct the works of man when they interfere. I found that if I was quiet and looked carefully, I could get a feel for what was happening physically in the environment, what had happened, had changed, was natural, was not. At moments I would—and still do—get a funny, tingling feeling on my inner arms, especially in natural settings new to me. I feel like the natural environment is telling me how it functions. Strange, I know, but that's the way it feels. (Something must be going on with my family. My father, his brother, and my own brother were very good water diviners, using only a piece of wire. I didn't have that particular sixth sense, but I could balance a green stick on my hand and use its movement to guide me to the water. Dad talked to me about feeling the energy. It's amazing. You *can* feel the energy of the water below you.)

After two great years with the Zulus in that swampy bush, I returned to the University of Natal in my hometown of Pietermaritzburg and took a pure science degree in geology. For the geology honors I studied two river mouths and a collection of beaches. Arguably, these were the first studies of the beaches on the eastern coast of South Africa. Occasionally, I'd get that feeling of electricity in my arms. I knew this is what I wanted to do—coastal work—and I soon touched down in Baton Rouge, my first trip ever to the United States. In fact, it was my first trip outside Africa, and I can thank David Hobday, whom I was extremely lucky to have as my senior lecturer. A tall, lanky geologist and extremely enthusiastic, Dave and I became friends. Once I decided to study overseas I sought his advice. He had done his Ph.D. at LSU and had some friends on the faculty who, as it happened, were coming out to South Africa for a conference. One of them was Harry Roberts of the Coastal Studies Institute; on the third day of our safari in the South African bush Harry invited me to work under him at LSU. Six months later, here I was, technically a "nonmatriculate" obliged to live in a dormitory for a year and obtain a 3.5 grade point average for a semester of senior-level undergraduate courses. If successful, I could then go forward with official graduate studies. At the time, LSU did not seem to recognize that other countries outside of Europe and North America have good education systems. In fact, those qualifying courses were a breeze for me, even a joke.

The Atchafalaya Delta was basically a wilderness and, for the most part, still is. It's rarely visited by anyone other than scientists, duck hunters, and bird watchers. In those days we had no radios and no safety flares, though I think we did carry some life preservers. It didn't matter to me. I'd spent so much time in the real bush, I had no qualms. (The alligators here don't come after you; compared with Nile crocodiles, they are truly pets. We did have to watch out for the bull sharks in the shallows.) My fieldwork called for a lot of heavy labor, literally hauling stuff around, and so did the work of my fellow grad students, so we usually helped each other out. One of these students was Paul Kemp, who would become my colleague at the hurricane center.

For six years I worked on deciphering the evolution and growth of the Atchafalaya Delta—a wild wetlands setting, yes, but in certain areas adversely affected by oyster shell mining activities (for roadbed material) and navigation channel dredging. Here I was drawn into my first environmental controversy in the United States. I and others felt that the shell dredging was eating up these exciting new Mississippi delta lobes. Louisiana was losing wetlands, and here the Army Corps of Engineers and the state were allowing this dredging company to just eat them up faster than they were being formed. I tried to fight these mining dredgers, working with a fellow from the U.S. Fish and Wildlife Service and a Sierra Club member. One morning I was called into a midlevel manager's office at LSU and told my "activities" were hurting my chances of getting my student visa renewed next time around. So I had to quit that campaign, but a month before returning to South Africa, with my Ph.D. firmly in hand, I got together with the folks fighting the shell mining and five months later received a newspaper clipping with the news that the shell mining had been stopped. There is always more than one way to cook a duck.

I probably spent three months out of every year in the wetlands. By the time I had my doctorate, I knew coastal Louisiana very well.

Just about the whole state of Louisiana south of Interstate 10 is new land, relatively speaking. Five thousand years ago the coastline was about thirty miles north of where it is today. The site of New Orleans was on a string of offshore barrier islands. *All* of this new land is sedimentation from the Mississippi River, laid down over the five millennia from the almost annual spring floods of the big river. The floods and the resulting sedimentation were not uniform throughout the vast floodplain, and the active delta of the river meandered back and forth, back and forth from Vermilion Bay in the west to St. Bernard Parish on the east, a stretch of coastline 180 miles across. We call these deltas "loci of deposition," and the switching from one to another is the process that has dominated this section of the coast—and, before that, a much larger section of the continent—for millions of years.

The river changes course in the wetlands because the sedimentation in the active delta necessarily raises the elevation of that land as it extends seaward, until the slope of the river is so flat over such a distance that it finds a shorter, more efficient course to the sea.

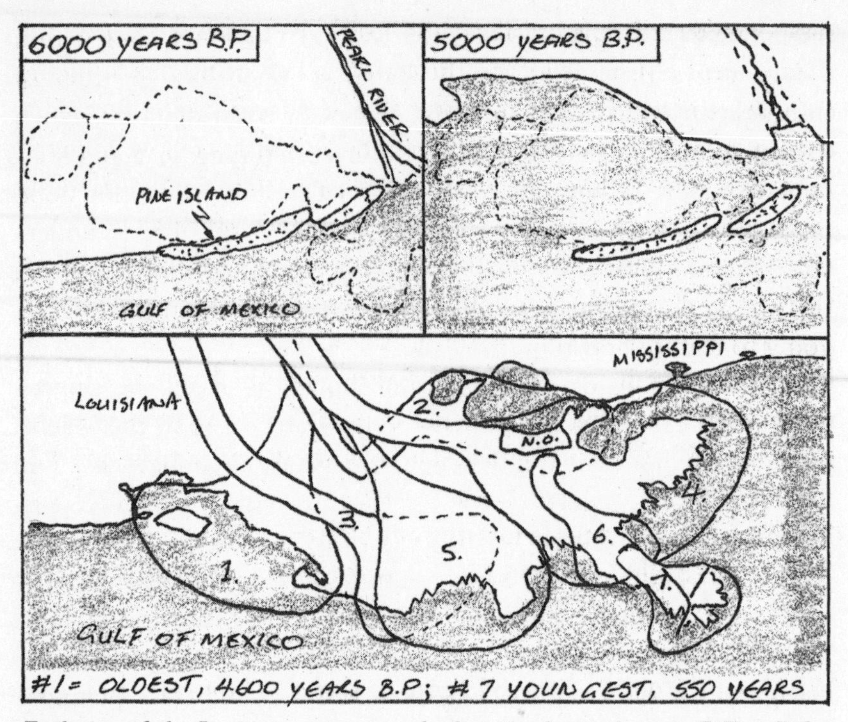

Evolution of the Louisiana coast over the last six thousand years (B.P. = Before Present). Note that New Orleans started off as a barrier island—Pine Island. This explains a lot of the beach sand that creates problems for levee builders.

Wonderfully adept science utilizing core samples and carbon-14 dating techniques has identified seven different deltas for the Mississippi in just the past seven thousand years. The geologists who specialize in this work have deciphered the layering for the whole coast.

The new Americans from lowland Europe who came to stay in this part of the continent brought with them a lot of experience with flooding and a good deal of know-how with levees. Most plantation owners built their own levee systems—with slave labor, of course. Founded after the floods of 1717 had subsided, New Orleans was built on the natural levee at the site, above the surrounding plain, and work immediately began on a mile-long, three-foot-high earthen levee on top of the natural one. This levee building continued for the next two centuries—and so did the floods. Following a terrible one in 1849, Congress passed the Swamp Land Act, and within a decade two thousand miles of levees confined long sections of the mighty Mississippi. In 1879, the Army Corps of Engineers, which had been created in 1802, got into levee building in a big way, but in support of navigation and channel training only, not flood protection—that was local government's role. Corps officials understood that the river and its levees would serve well as a jobs program that would never lose its mandate or its funding. And this work was going to feature brawn over brains.

In the long history of high water along this river, the tipping point was 1927. Before the awesome flood in that year, the federal government did not act under such emergency circumstances. The following year Congress directed the Army Corps of Engineers to build the system we have today. The whole story is the subject of a terrific best-selling book, *Rising Tide: The Great Mississippi Flood of 1927 and How It Changed America.* The author, my friend John Barry, also plays a significant role in following chapters about the investigation of the levee failures in New Orleans, and about planning for the future.

The latest generation of levees on the lower Mississippi River have done the job, virtually eliminating flooding in this stretch of the river. (The devastating flood of April to September 1993, one of the worst in U.S. history, was confined to the upper Mississippi.) They have entirely sealed off the river south of Baton Rouge. In fact, if it weren't for the determined efforts of the Corps, the Atchafalaya River, which breaks from the Mississippi two hundred miles upstream from the Gulf, would now be the main channel for the whole system, because it affords a significantly shorter course to the sea. But there was simply too much commercial infrastructure on the Mississippi below the junction with the Atchafalaya to allow that or any other distributary to take over the main flow and endanger navigation downstream. Also, the ensuing encroachment of saltwater up the channel of the Mississippi would endanger the freshwater supplies of many communities and industries. So the Corps harnessed the Atchafalaya at the Old River control structure, forty-five miles northwest of Baton Rouge, which opened in 1963 and now restricts the flow down this distributary to about one third of the total. (A big flood in 1973 almost washed away the original works at Old River. If it had, the Corps's best-laid plans would have been washed away as well, along with Morgan City and other communities downstream on the Atchafalaya, and that river would have taken over for good. Serious reinforcement was added to the original dam.)

The Mississippi River below Baton Rouge is a canal, really. It's a bit sad to think of it in those terms, but there you are. More to the point here—brutally to the point—this taming of the river is also the root of the problem for New Orleans and the rest of southeastern Louisiana. It is the main reason the entire region is so vulnerable to the catastrophic flooding caused by the storm surge of such hurricanes as Katrina.

To repeat, the entire Louisiana coastal zone is sediment from the Mississippi River deposited over millions of years. The sediments are usually highly organic, with lots of leaf litter and rootlets, and they have very high water content; as more and more sediment is added, any one particular layer gets squished from the new load overhead and therefore compresses and shrinks. That is, it sinks. Thus, marshland is just about the least stable of all soils. Imagine stiff yogurt, or soft playdough putty. It's not easy to walk across. Left absolutely alone, marshland will slowly compress under its own weight at the rate of a couple of feet per century. That's a lot, but there are locations in Louisiana where the soils are so mucky that subsidence becomes an issue within just a few decades—or, in some isolated cases, just a few years, or even faster in areas that have been drained for development. Yards in new subdivisions can sink within months of move-in, necessitating expensive yard soil replenishment.

The only way to compensate for the compression of the sediments is with ongoing, never-ending, inexorable sedimentation; otherwise, the wetlands will sink, and seawater will encroach and eventually kill the freshwater plants. The whole ecosystem will eventually collapse and sink even more. The whole process is more complex than this, of course, but this snapshot is a fair summation of what has been happening throughout southeast Louisiana for the past eighty years, simply because the levee system for the modern Mississippi River has prevented the flooding that yielded the natural accretion of sediment that created this soft land in the first place.

Or put it this way: Without new intervention, there won't be any land left to flood.

Local subsidence is not the only problem, however. Imagine the combined weight of all of the sediment that has been deposited by the Mississippi over all these millennia. The entire continental shelf sinks a little under this enormous load. Now match this wholesale systemwide subsidence with the fact that the seas are rising worldwide—six to eight inches in the last century, due mostly to long-term global climate changes, now due in turn mostly to our dependence on fossil fuels. Along the Eastern Seaboard, which has a small subsidence problem of its own, the local sea level has risen ten to twelve inches in the past century. In Louisiana the net effect is much worse, with the local sea level having risen on the order of thirty-six inches. And there is some good evidence that thanks to global climate change the sea level could rise at twice the current rate by 2100. This eventuality would doom much of the present shoreline around the globe and make the peril in Louisiana and New Orleans specifically even greater.

So the first point to know regarding the relationship of the disappearing wetlands and the peril in this part of the state is the irony that the flood-control measures necessary to create and then protect the infrastructure in this entire part of the state are contributing to the loss of the land on which this infrastructure sits. The second point is the impact of the oil and gas industry, without which the state of Louisiana would practically collapse, economically. Our wetlands are the nation's number-one source of crude oil (pumping more than the Alaska pipeline) and the second-leading source of natural gas, and in order to support and transport this production the companies have carved, by one calculation, eight thousand miles of cuts and canals throughout the wetlands. Since this entire network ties into the Gulf of Mexico, it provides opportunity for saltwater encroachment. It is subject to erosion and disrupts the natural flow of waters in the marshes. The whole artificial system works to the detriment of the wetlands. No one claims otherwise. The third undisputed factor is shoreline erosion, inevitable at all times but es-

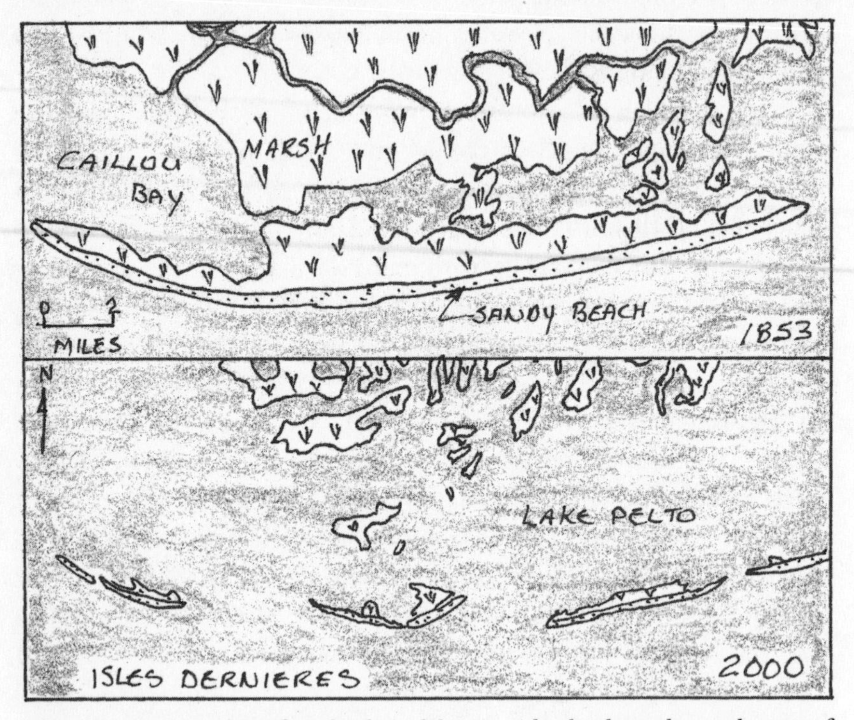

Note the dramatic loss of wetlands and barrier islands along the south coast of Louisiana, the Isles Dernieres chain. Loss of these wetlands means less surge protection; as a result some levee systems in the region are now too low.

pecially relevant when the natural sedimentation processes offer no compensation.

Just 150 years ago the ten miles between Isles Dernieres and the shoreline used to be a healthy marsh. It's all gone now, replaced by a shallow bay. In Cameron Parish on the western side of the state, Highway 82 has been moved back twice in the past 40 years, but it can't go back any farther, because it's on a ridge that protects a 30,000-acre marsh behind it. In the early 1990s the state set up 85 segmented breakwaters in a failed attempt to protect the stretch of coast between the beachfront communities of Holly Beach and

Constance Beach. Prior to Hurricane Rita sand had indeed built up on the eastern end of the stretch, but erosion has been serious to the west. Constance Beach used to have a two-hundred-foot beach and four east-west streets. Then it had almost no beach and just two streets. Following Rita, it is practically gone, as are most of the segmented breakwaters. Scientists had warned for years that the breakwaters would have no long-term positive impacts, would disrupt the local "alongshore" movement of sediment, and, when a hurricane struck, would all be wiped out anyway. All this proved true. Millions of taxpayers' dollars have been wasted.

The moral of this and many other stories in Louisiana: Know the science before you draw lines on a map and build hard structures. Serious consideration is now being given—and not just by bothersome scientists and officious bureaucrats, but also by the residents themselves—to deeding the lower coastal miles of Cameron Parish back to the Gulf of Mexico and moving everyone inland.

Many of the scientists who have studied the wetlands distribute responsibility for their disappearance more or less equally among the three factors: subsidence and the lack of river sediments, the oil and gas industry, and erosion. Without playing a numbers game, this seems to be fairly accurate. The infamous nutria, imported from South America for the fur market (reportedly, they escaped from their cages after a hurricane), are gluttons for marsh grasses and also blameworthy.

In the Crescent City, subsidence is both a natural and man-made phenomenon. The violent breaches of the 17th Street and London Avenue canals scoured deep holes in the soil and exposed the giant root masses of the cypress tress that thrived there up to three hundred years ago. The fact that these roots have not yet decomposed proves dramatically the resistance of cypress wood to water. When the engineer A. Baldwin Wood designed huge pumps powerful enough to drain the swamps between the Vieux Carre—the French Quarter—and Lake Pontchartrain, the city grew inexorably toward that lake until it reached the water itself. But the draining of the swamps also accelerated local subsidence dramatically. By definition, draining forcibly removes water from the soil, depleting its substance. The removal of the water also aerates the soil with oxygen, thus aiding the decomposition of the organic matter, depleting the substance of the sediment even more. Builders in the newly drained swampy areas understood the subsidence issues and knew that all homes had to be built on pilings, sunk seventy feet deep in some instances, so floors and walls would not crack as the soil sank unevenly beneath them. Of course, this sinking soil was the yard and many homeowners were dismayed to learn about the necessity of trucking in topsoil to make up for subsidence that could measure a foot a year. Moreover, the drainage canals dug to handle the floodwaters had to be (and still have to be) pumped regularly because they take in water from seepage. This water is from beneath the city. Thus the pumping also accelerates subsidence. The net effect is that the peat on which the city of New Orleans is built is compacting at a much higher rate than even the natural wetlands outside the city.

In recent years, two other, much disputed, theories have introduced new factors to be considered. In 2002, geologist par excellance Bob Morton, now with the U.S. Geological Survey, correlated the highest rate of wetland loss—thirty-eight square miles per year, in the 1970s—with the period of highest oil and gas production—those same seventies. Prior to Morton's work geologists had assumed that the oil and gas deposits were too far below the surface—eleven thousand feet to eighteen thousand feet—to have an impact on the rate of subsidence, but Morton hypothesizes that "regional depressurization" caused by the extraction of oil and gas is indeed a factor. Not surprisingly, the industry disputes his theory. If Morton is correct, there's nothing we can do about this part of the problem, because the oil and gas extraction is not going to stop. We'll be living in houseboats in Baton Rouge before that will happen. The other provocative theory is offered by consulting coastal scientist Woody Gagliano, who suggests that the large-scale subsidence of fault blocks—large sections of the earth's crust—is a major contributing factor, and there is growing evidence to support this idea.

The bottom line is clear: We are losing our wetlands in Louisiana,

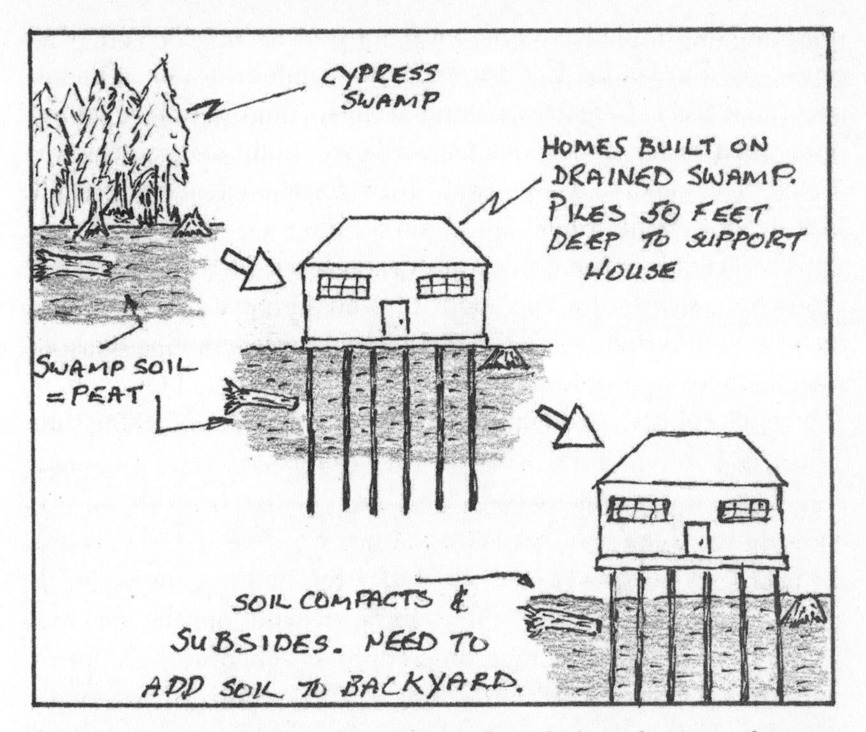

The soil subsidence problem in New Orleans. Recently drained swamp soils compact and subside rapidly, necessitating trucking in soil to maintain yards.

with dire consequences, of which the Katrina catastrophe is just the latest manifestation. Overhead photographs from half a century ago show many rich, healthy marshes and marshy prairies with small patches of water. Satellite photos of many of those same landscapes today show wide prairie lakes dotted with small patches of land. A fatal difference. If the marshes and swamps that comprise the wetlands are viewed as a sponge—and they often are in media stories—this one is in bad shape, ragged, with big holes and pieces hanging off. In the old days settlements out in the marshes were, for obvious reasons, built on the natural ridges and cheniers left behind as the whole sediment structure shifted over the decades and centuries. Those ridges might have had an elevation of eight feet to fifteen feet. Now three feet is more likely, because they're sinking. I think every major wetlands story in the press over the past decade has featured a colorful old-timer standing waist-deep in water where his home used to be, his face pensive and his mind "awash in memories," as *National Geographic* phrased it nicely. The old-timer's parents' house had been probably one hundred yards away, his grandparents' house five hundred yards away. Now those wetlands are totally wet—that is, underwater.

The average beach erodes at a rate of a few feet a year. At Port Fourchon, sixty miles due south of New Orleans and one of the main staging grounds for the vast oil and gas operations offshore, the beach loses forty feet a year. Highway 1 into Port Fourchon, on which tens of trucks roll up and down every day, floods all the time.

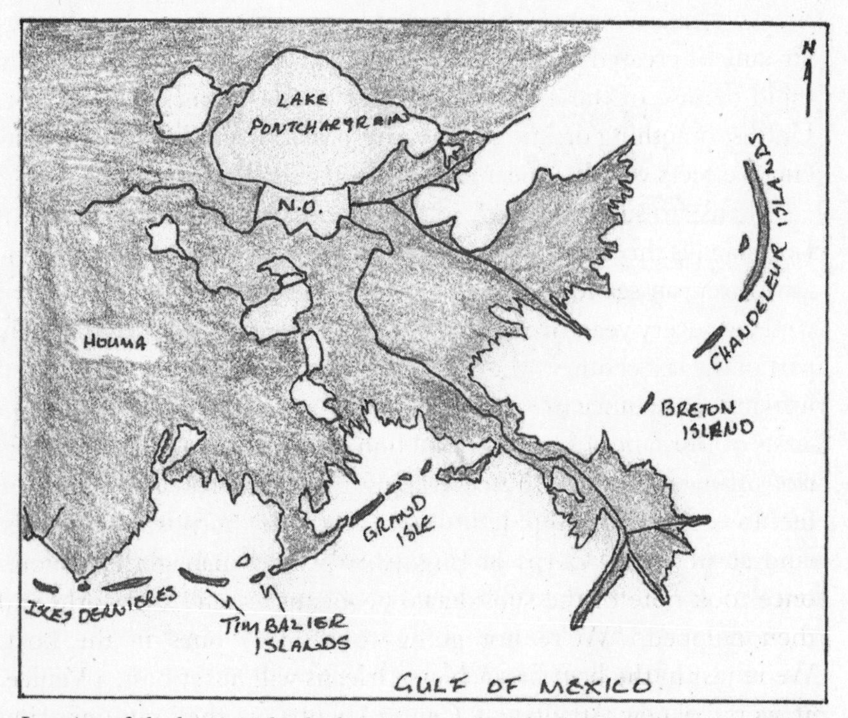

Location of the barrier islands that are crucial to the health and protection of the coastal region.

Any kind of southerly breeze sustained for a few days will do the trick. Throughout this part of the state, pipelines that were buried thirty years ago are now high and dry. The photographs are a staple of the magazines. It doesn't take a direct hit from a storm like Katrina to put the one-hundred-square-mile bird's foot delta (as we call it; there's a certain resemblance) in the very southeastern corner of the state—the mouth of the main shipping channel of the Mississippi—completely underwater. A tropical storm of any significance at all can accomplish this. (Ironically, the depth of the floodwaters caused by storm surge in some areas right on the coast is lower than in the past. Why? With less land to stop the water it just keeps flowing inland, inundating a greater area, some of it to less depth.)

We are also losing the protection of the aptly named barrier islands. Our activities have radically altered the natural sediment sources and pathways that maintained the barrier islands. Canals, breakwaters, segmented barriers, and other features of the new environment created by the oil and gas industry have all resulted in a rapid demise of these unique features of the Louisiana landscape. Unless something drastic is done, these islands and their many beneficial effects will disappear entirely in ten or fifteen years.

Thanks to subsiding land and rising oceans, the southern part of Louisiana is three feet lower than it was one hundred years ago, relative to mean sea level. The state loses twenty-five square miles of wetlands every year, over twenty times the rate recorded in the early part of the last century, when wetlands gain for the most part equaled wetlands loss. Since the 1930s, this has come to more than a million acres of lost land. In another one hundred years, if we don't get serious about rebuilding these wetlands, the land will be another three feet lower and, for all practical (and aesthetic) purposes, no longer land at all. Army Corps of Engineers project manager Al Naomi once took note of the subsidence problems around the world and then quipped, "We're not going to be only ones in the boat. We're just in the boat first." New Orleans will be, at best, a Venice, at worst, a new Atlantis—a Cajun Atlantis, as they say (ignoring or confusing the fact that New Orleans is a Creole, not a Cajun, culture).

This loss of the wetlands is not theoretical, not disputed, and not endorsed by even the wackiest observer. There are those who say, "Too bad—also too late," but no one says, "Hooray." This is because wetlands loss is not a zero-sum game. There are no winners. Every single interest in this part of the state—economic, environmental, and cultural—stands to lose. Alone among the economic interests in this part of the state, shrimpers, oystermen, and others who depend on brackish marshes to support the life cycle of their catch do benefit from the loss of freshwater sedimentation and the resulting encroachment of saltwater. But this benefit is for the short term only, because shrimp, for example, need marshes as their nursery, not open ocean. In the first phase of encroachment, the saltwater does create new brackish marshes, but the inevitable erosion that follows eventually destroys those same marshes and nurseries. These folks

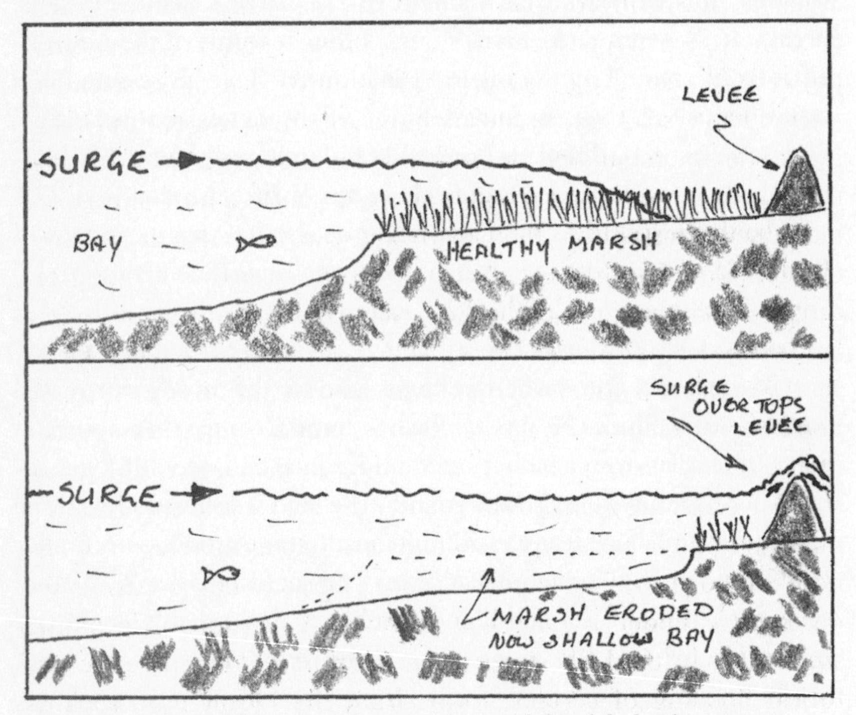

Healthy marshes swallow a lot more surge than eroded and depleted ones.

understand in their bones that the overall loss of the wetlands habitat is slowly, or not so slowly, eroding their way of life. They are often the men in the magazine photographs surveying the watery landscape that used to be the front yard.

Real estate developers hungry for terra firma are running out of acreage, so they dig canals to drain marshes and build ad hoc levees to protect the new sod from storm surge—until it doesn't. What companies would ever insure such homes? Louisiana and Texas traditionally rank one and two in flood insurance claims. Increasingly, insurers are not writing homeowner policies in the wetlands, or they attach an explicit rider excluding hurricane damage. I hate to vote with the insurance companies, but on this issue, who could blame them?

The wetlands can protect us from storm surge. Along with the barrier islands, they are the best, most natural, least expensive buffer available. Restoring the islands and the marshes is restoring lost friction, which we desperately need. Some scientists dispute the protective value of the wetlands. Some imply that other scientists (including myself) use the storm-surge argument as politically correct cover for rebuilding the wetlands for their own sake. It's true that I believe we should rebuild the wetlands for a host of reasons, but even scientists from the Corps have calculated that every three to four miles of healthy marsh reduces storm surge by one foot. (Recent research from the LSU Hurricane Center confirms this number.) Camille was a vicious Cat 5 storm, one of the most powerful on record in the Atlantic Basin, but areas that did not flood in 1969 *did* flood during Georges, a smaller storm that followed a very similar track a quarter century later. A more modest than expected Katrina still shoved ashore a record storm surge partly because only battered marshes stood in her way. One of the most important research questions we hope to resolve post-Katrina is exactly how wetlands reduce surge under a range of hurricane conditions. Upcoming simulations by the LSU storm-surge warriors will compare the historical flooding of certain storms from the recent past, such as Betsy, with the flooding that would ensue from that same storm today. In every instance, I'm sure, vast new regions would be underwater. The reason? The missing marshes and barrier islands. Joe Suhayda has already published computer studies showing that the 9.3-foot surge at Cocodrie during Hurricane Andrew in 1992 would have been a foot higher without the barrier islands. Joe has calculated that a completely healthy marsh system could cut the storm surge in New Orleans *by half.* Joe has always done good applied science, so I won't be surprised if ADCIRC corroborates those findings. I'll be very surprised if it doesn't.

Of the one million acres of wetlands that have been lost over the past decades, more than one half lie directly seaward of New Orleans and the one million–plus residents of the greater metropolitan area. If nature had her way, these wetlands would be sapping the strength of the storm winds and quite literally swallowing the brunt of the storm surges. This is why, as we rebuild New Orleans from the moldy flooring of the flooded city, it is necessary to also get serious sixty miles away. Rebuilding and restoration have to go hand in hand. That's my battle cry.

But how do we accomplish this? I'll get into the specifics in the final chapter. In this one I need to lay out briefly how the political establishment in the state has failed over the past several decades to rebuild the wetlands, and thereby left southeastern Louisiana in the position of the sitting duck for big storm surges. As we now know.

The undisputed relationship between lost wetlands and increased vulnerability to storms isn't new. The first alarms were raised shortly after I arrived on these shores in 1979, and they've been raised continually and at ever higher decibels over the past quarter century—the daunting challenge has always overwhelmed the political system's ability to generate a meaningful response. Maybe the deaths of more than thirteen hundred Americans will change the political landscape. Maybe. Based on my experience within that system in the 1990s, I have my doubts. After I completed my six years of postgraduate work at LSU in 1983, I returned to South Africa, worked on environmental management and restoration projects around the coast, including the response to Cyclone Demoina in Zululand, and then took over an underwater diamond-mining operation off the west coast of South Africa and Namibia. That was a fascinating experience, but the operation was sold when the owner decided to liquidate assets. With the political climate in the country deteriorating rapidly, I decided to sail the oceans, and ended up studying the coral reefs off St. Croix. In 1990 I flew from the islands to New Orleans for Christmas and ran into my old swamp mate, Paul Kemp, who told me about the new coastal restoration program that was just getting going in Louisiana, officially titled the Coastal Wetlands Planning, Protection and Restoration Act, sometimes also known as the Breaux Act, for former senator John Breaux, but known in the trade by its acronym, CWPPRA (pronounced "Quipra"). Congress had passed this legislation the year before and funded it for a decade. Also, the first Bush administration had put in place a "no net loss of wetlands" rule. If you destroyed marshes in order to build another development, you had to set up some kind of mitigation elsewhere.

"You should come back to Louisiana," Paul said. "This is right up your alley." I agreed. Restoration was right up my alley, and I hadn't heard about the new initiative. As much as I'd loved the wetlands as a graduate student, I'd pretty much lost touch with the situation in Louisiana. But now I'd turned forty, and somewhere in magaritavilla an Aussie had warned that I'd lose my edge, sciencewise, if I continued sailing for five years before going back to my profession. (That was my plan; five years was the estimated time needed to circumnavigate the globe, with some time lost to stocking up the cash kitty with odd temporary jobs here and there.) The Aussie was right, and six months after my conversation with Paul I sailed the *Maggie* into her new home waters, Lake Pontchartrain. Quite a change from island life in the Caribbean. Soon I hooked up with the consulting company where Paul had worked for a while, pushed hard on restoration activities, then ran into another former graduate student pal working with the federal National Marine Fisheries Service, and I set to work designing projects with them. Shortly thereafter, I moved over to set up LSU's Natural System Management and Engineering Program, an applied research program. Through it all I got to know well the CWPPRA program for restoring the wetlands.

It is impossible to understand how CWPPRA—or anything else concerning the environment in Louisiana—works without appreciating how the New Orleans district of the Army Corps of Engineers works, because it manages the money and a large share of the projects. With an Army colonel at the top as commander and district engineer (some are not even engineers), the agency has a military-management veneer, but the real power lies in the civilian career managers. For one thing, every three years the Corps rotates the district engineer, so the learning curve at the top starts all over again, and three years might easily be required to move a project from a priority list to construction. Since I returned to Louisiana, there must have been half a dozen chiefs. I can't even recall all their names. Power necessarily devolves downward; the civilian managers really control both the short- and long-term agendas. Because they are long term and folks like Paul Kemp and myself are also long term, we give these managers fits. (The same folks in charge of the design and building of the levees that failed during Katrina are now fighting our Team Louisiana levee investigation, as we'll see, and have convinced their nominal superiors to spend a considerable amount of money on their own investigative teams.)

Across the whole country, most of the Corps's funding is for "earmarked" projects requested by specific members of Congress. Political interference, we call that. Pork. CWPPRA has the same problem in Louisiana. A task force bringing together five federal agencies and the governor's Office of Coastal Activities approves all projects—a total of 141 so far, including 62 completed and another 19 deauthorized. Some projects have been approved but not funded, others funded but then never begun. (The money comes from a user fee on small engines, boat fuel, and certain equipment.) In addition to the Corps, the federal agencies in the task force are the Fish and Wildlife Service, National Marine Fisheries Service, Natural Resource Conservation Service (formally Soil Conservation Service), and the Environmental Protection Agency. If each receives what it considers a fair share of the funded projects each year, it can maintain its workforce for the next year as well as keep its constituents happy. Each agency also has its special interests and expertise. Wildlife people tend to be concerned about habitat; soils people tend to be concerned about marsh management; and the EPA and

National Marine Fisheries are concerned about wetland creation. The oversight and funding mechanism was never an ideal mechanism, but what can we do? This is the nature of human beings and our bureaucracies.

There also seems to be an inherent suspicion between the Army Corps of Engineers and scientists, who in turn tend to believe that the Corps automatically advocates "hard" structures because these are what the Corps builds. For their part, Corps officials know that their concrete has saved the infrastructure in this entire part of the world from river flooding for the past half century. They understand the downside, of course, but they're not going to apologize for taming the Mississippi.

In the first decade of the CWPPRA program, Louisiana received about $40 million annually from the fund and added $5 million more from the Wetlands Trust Fund, which receives royalty money from the oil and gas industry, and the state also receives some in-kind payments. Several hundred million dollars are in the bank awaiting disbursement for projects funded but not yet begun or completed. The great majority of the projects are small, sometimes demonstration-level, and can be slotted into one of three categories: shoreline protection, wetlands creation, or hydrologic restoration. Shoreline protection is just that, usually accomplished by the placement of rocks. Wetlands creation is also just that. Sediment is mined and placed as new "proto" wetlands; dredge material from navigation channels is used beneficially, or river diversions bring in sediment-laden water from the Mississippi River. Hydrologic restoration generally involves manipulating water levels within marshes, including "marsh management" projects in which a levee or dike is built around a marsh to artificially lower the water level in order to sustain freshwater plants. Such management projects are not strongly supported by most coastal scientists. I could argue that they principally benefit ducks—and duck hunters, a major constituency in Louisiana. The very savvy Shea Penland, a fellow LSU graduate who runs the fine coastal research program at the University of New Orleans, has studied these projects carefully and concluded that more wetlands loss has occurred within marsh management areas than outside of them. *Ten times more*, if you can believe that—which I can. Wetlands creation is the way to go.

The CWPPRA program has been controversial from the beginning, often because this or that part of the state is allegedly getting more than its fair share of the money. (See "Coast in Peril," environmental reporter Mike Dunne's superb account of this issue and the whole wetlands history published by the *Baton Rouge Advocate* in 1999.) Sometimes the CWPPRA task force has found itself in competition with outside developers, most notably with its plan to restore over twenty-five hundred acres of marshland at Eden Isles in St. Tammany Parish, on the north shore of Lake Pontchartrain. Before it could restore this land the task force had to buy it, but it was outbid—twice—by housing developers. So this project was eventually deauthorized and defunded. We have instead today yet another major upscale housing development, with canals dredged to raise the elevations, and all canals connected to Lake Pontchartrain. Instead of better surge protection for Slidell, we have more infrastructure for the higher surge to damage. If these twenty-five hundred acres had been restored, the marshes would have helped during Katrina, cutting down the surge that hit the eastern part of Slidell.

Nevertheless, in the beginning, I was really excited about the opportunity to launch some first-class environmental restoration projects. Soon enough, however, reality set in, as it usually does, and not just with the little squabbles about the basin that got 7 percent too much money. I looked around the table and wondered, where is the thorough understanding of the science here? The CWPPRA legislation had been drawn up by scientists, environmentalists, and academics avowedly intent on bringing the science to bear, to provide checks and balances with the other interests. But at one of my first CWPPRA meetings, an important restoration official at the Department of Natural Resources (DNR), said we did not need any more science or any academics. I was astounded! *All* we needed was science—and money—and professional, apolitical management. Unite these three and step aside! That's what we needed to do, and do it with care, because when we change a complex physical environment, we change everything.

The analogy I sometimes used to make my point was a simple tank with some water and some fish. Start adding and subtracting water. Nothing much changes. But now fill one side of the tank with soil that forms a gentle bank sloping into the water. Now we have both water and land, and, in addition to the fish, a few plants and a little animal or two. Any audience can intuitively sense that the simple addition and subtraction of water and soil to this more complex environment may yield not so simple consequences for every element of the story, including the fish and the animals. In a dynamic environment like coastal Louisiana we have to understand these processes, because the details of the physical environment drive the system. With these audiences I was also subtly trying to sell the fact that we need sediment, sediment, and more sediment—correctly distributed—if we are to build wetlands and protect people.

On a related front, Paul Kemp and I were also battling the method used to evaluate potential CWPPRA projects, something called the Wetland Value Assessment, which was sans science and stank to high heaven. This scheme was devised by wildlife experts with no real understanding of the dynamics of the overall system. It was heavily weighted toward marsh management projects. A profound deficiency was its failure to give any credit whatsoever in the cost-benefit analysis for the value of wetlands as storm-surge buffers. What about Hurricane Andrew? Earlier I cited the important number: Thanks to the relatively healthy wetlands along the central part of the coastline where this Cat 3 came ashore in August 1992, the storm surge was only eight feet.

Another profound deficiency was the back-of-the-hand dismissal of the barrier islands, which got no credit for their contribution to surge buffering. They were not really considered as part of the wetlands "package." Other than the EPA and National Marine Fisheries—always the progressive groups—the other agencies were dead set against restoring the barrier islands. I was told a hundred times that the "W" in CWPPRA was for "Wetlands." Did I see a

"B" or an "I" in the acronym? So we could restore only *wetlands*, not barrier islands. Oh, my. For shortsighted bureaucratic thinking, I give that an "A."

In my earlier discussion of the ADCIRC storm-surge model I told the story of how Paul Kemp, Joe Suhayda, and I convinced Terrebonne Parish to give us thirty thousand dollars to do some early modeling to compare what the tides did to the marshes south of Houma, with and without the barrier islands offshore. The results were, as I said, irrefutable. Rebuilding the islands to their state one hundred years earlier would immediately stop the flooding of thirty thousand acres; "wetting times" would be much lower throughout the basin, very beneficial for wetlands trying their best not to drown. I've described the reactionary response that greeted this model in meetings. If I showed some of the CWPPRA folks what they wrote back then about the computer models, I'm sure they would be quite embarrassed.

The missing science was one problem that had to be addressed. Another, just as pressing, was the missing big picture. To my mind the most important part of the CWPPRA legislation was the long-term comprehensive planning effort, but, in fact, the larger and therefore probably more controversial projects lost out to smaller, straightforward, maybe even demonstration-level jobs that disbursed the available money without too many hassles. In November 1992 I organized a workshop in Baton Rouge, pulled in as many scientists as possible, supplied sandwiches, and explained the prototype version of our big-picture plan. We had to move away from the laundry list of small CWPPRA projects. These Band-Aids would never save the patient. No matter how many we used, we would, sooner or later, end up with a carcass. Of that I was sure, and so were many others. We had to turn things around with big, global thinking, with triage. In all of my classes I've always preached that we have to understand the evolutionary pathways of any given environment. How has man influenced it? How have we speeded up natural change, slowed it down, stopped it? For every movement there is a reaction. "Least interference" is the goal, but sometimes this goal requires major interference with the preceding interference in order to set things right. So it was and is with Louisiana's wetlands. Only large-scale diversions of the rivers and committed rebuilding of the barrier islands will do the job. Hurricane Andrew had grabbed my attention. I was surer than ever that we were sitting on a ticking time bomb. I wanted to organize these scientists into a united front, then take on the agencies.

We got the scientists on board—not difficult, the science is crystal clear—but had much less success pushing the comprehensive approach at the various meetings of the CWPPRA basin teams. (There were nine in all, each responsible for one basin in south Louisiana. I was the scientific adviser to the team in charge of the Atchafalaya and Mississippi River deltas. Such was my modest little platform.) Then, in mid-1993, Colonel Mike Diffley, district engineer for the Corps, called a meeting of all the teams. One after another the presenters laid out plans beholden to the usual approach. It was one small Band-Aid after the other. The colonel was uncomfortable with what he was hearing. I could see that and waited for my chance. (It's true that the real administrative power in the district seems to reside with the career civilians, but a boss who gets the big picture is still a tremendous asset, even though he'll be gone in three years, at most.) When Colonel Diffley called for a break and walked off to lean against the wall, away from everyone else, I approached him. As the old saying goes, "In for a penny, in for a pound." I told him that the academic/scientific community was going to shoot down these plans, loud and clear, because they showed no imagination and no grasp of the big picture. We had to change course in a fundamental way. As quickly as I could, I listed the four main bullet points of the big-picture approach. All the while he stared at the floor, then suddenly lifted his head and looked me right in the eyes. I assured him that we scientists would be more than willing to help. Just reach out to us at LSU and we would bring in the academic community and give him something he would be proud of. He knew how to find me, but I gave him my card anyway, and as he went back to his seat at the head table I said a quick prayer that I'd made a good case, and that he had heard me.

After the break the colonel stood up and graded the basin plans he had just heard D-, maybe only F+. The room was very silent. Everyone respected this man. What he said had impact. About a week later he came to LSU by himself, met with Paul Kemp, me, and one or two others, and asked us to write the introduction chapters for our plan, as well as the executive summary. He didn't have any money to give us, but he was accepting our offer of help. We had ten days. (Be careful what you pray for!) I don't know whether Colonel Diffley was just taking a what-the-hell long-shot gamble on a plan that might fall flat on its face, or whether he actually believed in us. Regardless, Paul and I tore into the task with twenty-two-hour days. In the end the job seemed worth it. Diffley was pleased and told us not to worry too much about the fact that the big-picture ideas were not yet fully developed. "All anyone will read will be the executive summary," he said, and I knew he was correct. Paul Kemp and I were elated, as were our supporters, but in the whole process we made some enemies in the Corps's civilian upper management, a fact of life that still haunts us to this day.

As one of the authors of the CWPPRA document, I decided to spice it up for a wider audience, add a great deal of detail, and publish it in 1994 as a white paper on the wetlands under the auspices of LSU's Center for Coastal, Energy and Environmental Resources (CCEER). The title was "A Long-Term, Comprehensive Management Plan for Coastal Louisiana to Ensure Sustainable Biological Productivity, Economic Growth, and the Continued Existence of Its Unique Culture and Heritage," but it became known (of necessity) as the CCEER report. The reference to "economic growth" was not a second thought, nor was the reference to culture and heritage. Along with my conviction that a big picture guiding big projects was the only salvation for the wetlands, I had developed a second one: that the politics could be put together only if restoring the wetlands and the barrier islands was primarily for the benefit of *people*, not of herons and alligators and marsh grasses. Well aware that public support would erode if restoring the wetlands was believed to be transgressing too much on property rights, I acknowledged economic dislocations but denied net economic losses. On the contrary, rebuilding the wetlands will yield major economic gains to the fisheries, protect the oil-and-gas infrastructure, and build ecotourism. (Before Katrina a relative handful of New Orleans's millions of visitors took an air boat out into the swamps and marshes. That market is a potential gold mine.)

The center received hundreds of requests for copies. The new publication was well reported in the media. I got congratulatory letters from far and wide, including from Senator Breaux. The agencies were a different story; some that are friendly now weren't so friendly then, but so what. I believe this paper remains an excellent guide post-Katrina—more than that, a mandatory element of the overall rebuilding effort, and I will lay it out in the final chapter.

In July 1994—about two thirds through his second term—Governor Edwin Edwards appointed a new head of the Department of Natural Resources, Jack McClanahan, an oil-and-gas man who immediately brought in another guy from the oil patch, Gene Spivey, to be his executive assistant. I knew I needed to meet the new chief of the most important of the state agencies associated with CWPPRA, and with the help of my friend Glenn Wood, a dredging operator, who was also a friend of McClanahan's, I got an interview with both him and Spivey. I give these two a lot of credit. They listened, they read the CCEER white paper, they endorsed the big picture. And to make the long story short, two weeks after meeting McClanahan he asked me to become the state's new restoration "czar." What an amazing change of fortunes for me. My predecessor, Dave Solieu, was the consummate insider. Here I was an academic with no political ties to the governor or anyone else whatsoever. How strange was that? I was about to get a whole different kind of education, one that has seriously colored how I now look at our long-term prospects in Louisiana.

The green forces in the state were confused and delighted by my appointment. Did Jack McClanahan, of all people, intend to cut the Gordian knot? No, I don't think so, but his stature as an oil man made it a whole lot easier to sell our "radical agenda"—that is, actually doing something big—as it would soon be labeled. A complicating factor was the ticking clock. I felt I had Edwards's support, and Jack McClanahan's, and I had the mandate to make changes, but Edwards was term-limited. He'd be out of office in January 1996, only fifteen months away. That might be the end for me as well, so we had to get something big in place; we had to convince everyone that we had a comprehensive, big-ticket plan, and we were going to see it through. In January 1995, we held a coastal summit and followed it with a white paper. We set out to reassess all projects so far authorized by the CWPPRA task force and to cut the smaller pork projects. The state wields something of a de facto veto over CWPPRA projects, because withholding our share of the cost effectively kills the job. So don't even bother putting up any bad ones for consideration. The word spread, and it really rang the bell hard.

The basic idea was to start rebuilding the barrier islands first, then to build the wetlands in the protected bays behind these. Immediately, we set to work on the project to mine the sand from Ship Shoal eleven miles offshore, using it to restore the islands. The science here was straightforward, but the politics were not. I spent a lot of time in Washington, working with the congressional staffs and the appropriations committee staffs. We got the Louisiana delegation to write letters of support. We got the very important congressman Billy Tauzin on our side. Tauzin told some folks from Houma that I was now "one of them." He took me to meet Congressman Bob Livingston, who in turn advised me to get every national environmental group on board. Along with Mark Davis, the new director of the Coalition to Restore Coastal Louisiana, I briefed as many of these groups as I could. I got $400,000 for the impact study from CWPPRA, with the Mineral Management Service "tasked" to do the job, which made sense, because this agency controlled, as a federal resource, the sands in question. By the fall of 1995, therefore, we thought we had all our ducks in a row. The Corps said they'd sign off on the project, given a proper feasibility study. Just about everyone at the state level was on board. We had the commitment to deauthorize a bunch of projects that had no value. We had the commitment to use one third of all CWPPRA money for big-ticket projects. We had letters of support from most of the congressional delegation. We had the national environmental groups on board, as long as we had the detailed impact study. Things looked good—so I quit my job as coastal restoration chief.

The timing seems odd? The explanation begins the previous year, 1994, when representatives Billy Tauzin and Jimmy Hayes had introduced the Private Property Owners' Bill of Rights. This legislation would have directed the government to compensate landowners when environmental regulations deprive them of 50 percent or more of the fair market value of their property. This one-sentence description of the bill sounds fair enough, but in reality it would have allowed you, a landowner, to claim that you would certainly have built a Wal-Mart on your one hundred acres of marshland if it weren't for these communistic environmental regulations, so you want full compensation for your lost fortune. The legislation also reclassified wetlands on an A, B, C, and D scale in terms of importance, and most would have been D—unimportant. The attorneys general from thirty-three states and territories wrote Congress to oppose this bill because it "would write into law the dubious principle that the government must pay polluters not to pollute." Simply put, this was narrow-minded, shortsighted, knee-jerk so-called property rights legislation. If every significant restoration initiative has to fight a horde of one-hundred-acre landowners blackmailing the federal government for maximum "Wal-Mart" compensation, forget saving south Louisiana for future generations. Such mind-boggling obstructionism would have killed coastal restoration, period, and guaranteed one Katrina catastrophe after another, until there was nothing left around here.

As this legislation made the rounds, we felt compelled to write letters to our congressional delegation making clear exactly what this legislation would mean for the long-term future of their state. The letters went out in July 1995. Contact us if you wish to discuss this further. We waited two weeks and heard nothing from any of them. Not one word, so Jack McClanahan, my boss, told me to release a copy. The media loved this letter, of course. Here was the oil man who headed the Louisiana Department of Natural Resources and his coastal restoration guy (me) saying that Billy Tauzin's legislation was wrong. There is no doubt that the letter did make Tauzin and Hayes look bad, but how else were we to oppose their incredible bill? We knew we had it right, and public opinion polls consistently showed public support for our work.

One day shortly after the letter was released, I was back in Baton Rouge enjoying a po' boy at a local establishment when I got a call from Tauzin himself, swearing, labeling me a "tree-hugger," etcetera etcetera. Then, at a CWPPRA meeting that Senator Breaux called in Washington one month later, Tauzin and Hayes launched into me. I was an "African dictator," and worse. Hayes said he would get me. Funny enough, one of Tauzin's staffers had told us a few months before that his boss was impressed with the progress we were making. That credit was totally irrelevant now. What should I do? The gubernatorial election was coming up in a few months, and Governor Edwards would be out of office a couple of months after that. I could see that folks were lining up for personal favors, and as the assistant secretary of the DNR, I was in charge of the coastal zone permitting division. I was wary of getting involved in the last-minute permitting deals that are almost a tradition in Louisiana and elsewhere, I'm sure. (Years later, an FBI agent who was part of the team that sent Governor Edwards to the penitentiary told me my instincts in that regard had been good.) Two powerful congressmen had put out a hit on me, politically speaking. Adding all of this up, taking into account my feeling that the big-picture initiative was in good shape and moving ahead, that I had accomplished what I had set out to do (Jack McClanahan had even received the EPA's Wetlands Award for 1995), I resigned.

This was a sad parting, definitely, but until McClanahan and Gene Spivey left when Mike Foster's new administration took over in January 1996, I was in constant contact with them and, in some ways, the acting restoration chief. I remained optimistic. Then everything suddenly collapsed. *Everything*. With Edwards out and Foster in as the new governor, with Jack Caldwell taking over from Jack McClanahan as the new secretary of the Department of Natural Resources, the prospects for responsible coastal restoration fell apart within two or three months. Caldwell's staffers sent letters to the state team working on the Ship Shoal feasibility study, questioning if there was any value in restoring the barrier islands. The dynamics of the feasibility study changed, and the funding for the Mineral Management Service part of the investigation was removed. Those folks were just "chased from the table," as one of them told me later.

Unbelievable—the wake-up call of a lifetime for me, and a blunt lesson about the vindictiveness of Louisiana politics. The Ship Shoal mining project was dead, along with the comprehensive, big-picture approach to solving the wetlands problem, all because of petty politics from the junior bureaucrats who were influencing the new man. The agencies that were used to getting their little share of CWPPRA funds every year wanted their little share every year, and they won out. I had no idea that these desk jockeys had the wherewithal to sabotage us.

Such was my initial postmortem. In fact, though, the linchpin may well have been my falling out with Billy Tauzin and Jimmy Hayes over their ill-conceived takings legislation the previous summer that had died the ugly death it deserved. They had it in for me. When the administration in Baton Rouge changed, the two congressmen got their pound of flesh from me and more when they helped kill our comprehensive restoration initiative. Subsequently, a senior LSU employee told me that Tauzin's staff had also warned the university that if my name was on anything that needed funding—forget it. This whole episode proved to me that politics in Louisiana was getting in the way of restoration. I had seen how quickly the political system in the state, with just a change of governors, could totally derail a long-term restoration program that we had thought was cast in stone. (I guess I can be *very* naive.) The state therefore blew a golden opportunity to do something great. I saw immature, petty, childish, and pathetic tit-for-tat politics as the order of the day, and I saw that CWPPRA was not going to restore the wetlands, the coastline, the barrier islands—anything at all.

All in all, my tenure as the restoration czar was terribly disappointing regarding the big picture but gratifying regarding some smaller accomplishments. It was an incredible learning experience. Now, what would be the next best thing to do? I cast around for the new right focus for the long haul. As noted, I had always been impressed with Joe Suhayda's early storm-surge models. Now he had flood models for New Orleans. This was the future, I knew. Once we could really prove the importance of the wetlands in minimizing storm surges and protecting people, coastal restoration could live again. Hurricanes—storm-surge models—coastal restoration. They all had to come together. I also realized that we knew absolutely nothing about the actual impacts of hurricanes—specifically, about the impact of a major storm on southeast Louisiana and New Orleans. I started thinking in these terms, and this was the train of thought that would lead to the creation of the two hurricane centers at LSU, as I've related.

In 1998, right around the time of Hurricane Georges (how many wake-up calls did the people and their politicians need?), a new task force impaneled by Governor Foster estimated that current spending from all sources (mainly CWPPRA, of course) would mitigate only 14 percent of the projected land loss by the year 2050. Even given perfect results for every project, the state would still lose six hundred square miles of marsh and four hundred of swamp land. After dozens of meetings around the state, yet another new blueprint was produced, this one labeled "Coast 2050: Toward a Sustainable Coastal Louisiana." It called for spending $14 billion over the next twenty to thirty years, a threefold increase over what could have been expected from the CWPPRA schedule. But it was still mainly Band-Aids, and I failed to see how more of them would save this patient, who was in extremis. I was joined in my criticism by others. In our opinion, the only answer remained what it has always been: big, big projects.

I was excluded from the "Coast 2050" initiative. The plan referred to my previous planning efforts, but I was never asked to this new table. Maybe they knew I would be skeptical of Band-Aid approaches, but I imagine it was mostly just the dirty game of politics as it's played by so many in Louisiana. "Coast 2050" was then followed in rapid succession by a dizzying array of other, overlapping studies, plans, and funding schemes, none of them ever likely to produce the kind of money or vision required. To be frank, I lost track of the fate of some of them—in the particulars, that is. Overall, nothing much happened.

In 2000, Florida got $4 billion to restore the Everglades, matching it with $4 billion of its own funds. President Clinton officially launched a historic thirty-year restoration, supported by a broad coalition of national and local environmental, agricultural, business, and citizen groups. Someone quipped that our wetlands make the Everglades look like a petting zoo. That's correct. Why couldn't Louisiana coax that kind of money out of Washington? Pretty soon, Billy Tauzin, who had gotten so upset with me—and vindictive—for opposing his takings legislation, got religion of a sort and began exploring the possibility of funding the dust-gathering "Coast 2050" initiative from the Water Resources Development Act, under which the Everglades project was funded. As chairman of the House Energy and Commerce Committee, Tauzin was generally considered a point man for regulated industries, although oil and gas companies understand better than anyone the jeopardy of their inland infrastructure. They'd be delighted to see federal money rebuilding the wetlands that their own activities have been significantly responsible for tearing up. (Shell was a big backer of the state's America's Wetland public relations campaign kicked off in 2002. So was the maker of Tabasco sauce, headquartered at Avery Island.)

Quite a few people have started getting religion on the wetlands in recent years, at least for political purposes. We've been hearing less poppycock about "people first" and "people instead of wetlands." This is not and never has been an either-or issue. It is and always has been people *and* wetlands. Tauzin's new idea was not terrible, but the story was still not quite right. It needed the hurricane threat, the storm-surge threat, the imperiled people.

The environmental impact plan for Tauzin's idea has been submitted, but of course all bets are now off. The congressman himself has retired and taken a new job as head lobbyist for the pharmaceutical industry, over which his old House committee had jurisdiction. So it goes. Just before leaving office he made a heartfelt plea before the House Transportation and Infrastructure Committee. "Our paradise is about to be lost," he said. "Please let's don't have a commission where all of us, red-faced, say we saw it coming and didn't do anything. Please don't let that happen."

I now propose a moment of silence.

NINE

THE SCORE ON THE CORPS

On Saturday afternoon, September 3—five days after Katrina's landfall—our paradise—part of it—was indeed underwater, and surge warrior Hassan Mashriqui sent around a note asking about the total capacity of the 148 pumps at the 60 pumping stations in the metro New Orleans area. (Twenty-four of the stations are in the two western bowls. Orleans East has eight; St Bernard, eight; and Orleans and Jefferson parishes south of the Mississippi, twenty.) Mashriqui had pinned down the volume of water in the Orleans Metro Bowl at thirty billion gallons. Now all he needed was pumping capacity and we could get a reasonably accurate idea on how long the unwatering (a word coined by the Corps, I believe) would take, once the pumps could be powered up from the grid or generators and operate at some reasonable percentage of capacity. Of course, that "reasonable percentage" was the catch, but we needed to do better than the unwatering estimates thrown out over previous three days, which had been all over the place and were therefore not very helpful. The mayor said one to two weeks—not necessarily absurd, but highly unlikely. Colonel Richard Wagenaar, the commander and district engineer for the Corps in New Orleans (newly assigned in July, transferred from Korea) estimated three to six months. One FEMA official estimated six months to drain, three more to dry. God help us.

Somehow Newt Gingrich, the deposed Speaker of the House of Representatives, author of the Contract With America (or "On" America, depending on how you looked at it), now a fellow traveler with Senator Hillary Clinton on certain issues, got involved in a string of e-mails and suggested that there must be professionals in "drying things out" who could beat the current predictions. What about the Dutch? he asked. Someone else suggested contacting the dredging companies. Maybe they could position a long line of their barges along the Industrial Canal, say, and, instead of mud from the river bottoms and canals, pump out water from the bottom of the lake that was now New Orleans. I hope I didn't join in the wild numbers game—I don't believe so—but I did succumb to the pessimism when I told a reporter that the pumping issue was a "mission impossible" if they expected to achieve it within a couple of weeks. Looking back, I wonder whether the increasingly dire estimates for draining the water reflected not the best judgment but the deteriorating conditions in the city and the increasing sense that everything in this zone of devastation was going to hell. Pessimism sometimes seemed like the only reality.

The pumps for the 17th Street and London Avenue canals could not be turned on until the three breaches of those levees were patched effectively, because these pumps are deep inside the city. Engaging them with the breaches still open would simply pump the water back into the city by way of these breaches. Patching these breaches would be easier now that equalization between the lake and the water in the city had been reached, but the Corps was having a hard time finding contractors who could get their equipment into the city. Some of the pumping stations themselves were flooded and couldn't come on line until other pumps had lowered the water sufficiently. The debris problem, a given in a normal flood, would be exponentially worse in this one, because the water was so full of everything imaginable. Unless constantly cleared from the pump inlets, this debris will block the flow and place the pumps themselves at risk. Of course, the pumps require electrical power, of which there was almost none. Portable emergency generators were now on hand, but these required fuel, diesel or natural gas. Where were these replenishable supplies, and could the generators be positioned near a given pumping station? Someplace dry was required.

Another question was whether this water—this toxic stuff—

should be pumped directly into Lake Pontchartrain, from where it could spread into the marshes and swamps, or into the Mississippi River, thence into the Gulf of Mexico? The question was asked, but only two of the pumps for New Orleans discharge into the river. The priority had to be getting the water out of the city, no matter where it ended up or how toxic it turned out to be. The water would go into the lake.

At the LSU hurricane centers we had posited the pumping problem as a critical one in a flooded New Orleans, but we hadn't studied it "officially." It was our clear understanding from the Hurricane Pam exercise that, given an impending disastrous storm, the Corps would preposition barges loaded with pumps that would be ready to start the job. We had no reason as yet to second-guess the Corps, but on that first weekend after the storm, I was among the pessimists. Fortunately, some of the people charged with getting the pumps going were wonderfully ingenious. In a fine story on September 8, the *New York Times* would recount the saga at pumping station #6 located at City Park Avenue on the 17th Street Canal, the city's largest station, vitally important. In charge of the operation was Chief Warrant Officer Thomas Black, an engineer with Army's 249th Prime Power Battalion, recently returned from sixteen months in Iraq. A helicopter tour of the vicinity around the station had failed to pinpoint a suitable location for the emergency generator, but then someone noticed a stoplight working in adjacent Jefferson Parish. Could this team somehow link pumping station #6 to that electrical grid? Over two days they set up "a giant extension cord," and on Monday night, one week after Katrina hit, six of the fifteen pumps roared into action. (The others are so old they can't handle the standard 60-hertz current. They need their own 25-hertz generators.) The reporter noted that the racket could be heard from blocks away at night, since otherwise the city was dead quiet.

Mashriqui put everything together and concluded that the city could be empty in two to three weeks. That was quite an optimistic figure. In the end, the entire metro region—all of the bowls—would be declared officially dry on October 11, just over six weeks after landfall. This somewhat extended time frame reflects in part the reflooding that followed levee failures along the Industrial Canal during Hurricane Rita in late September.

Exactly how bad was this witch's brew of water? We definitely needed to get going on this answer, and on the first Saturday after the storm I joined John Pardue, director of the Louisiana Water Resources Research Institute and a member of our LSU team, and headed off to get what early samples we could from the lake and the city. The marina on the north shore from which we set forth was a smelly place: dead fish and decaying biota. Once on the water I switched on the VHF and immediately heard that the lake was officially closed. The fine for approaching New Orleans by water was fifty thousand dollars and the risk of imprisonment. That punishment didn't seem likely, but I could envision being greeted by soldiers with rifles. We could probably talk our way past them—we had all kinds of useful identification—but the simpler course of action was to get permission, and after many tries, I managed to get through on the cell phone to Kevin Robbins, who directs the LSU team at the EOC, and he immediately arranged clearance for us with the Coast Guard.

The causeway across the lake was open to emergency vehicles, and there was a fairly steady stream going both ways. Above the city helicopters were buzzing everywhere in serious numbers, for the first time. Five days after the storm thousands of people were still stranded on rooftops, overpasses—just about any high ground. At the Industrial Canal, where we sampled the water flowing out, John and I heard at least one shot fired. At the London Avenue Canal we watched employees of the Orleans Levee Board adding mud and stones to the seal they had built at the bridge, near the entrance to the lake. One well-armed state policeman was standing there, all by himself, and not very friendly. (The two breaches at this canal were surrounded by floodwaters and inaccessible to vehicles, unlike the one at the 17th Street Canal, so serious work had to wait until the middle of the following week.)

The Environmental Protection Agency would be all over the water contamination question, partially in response to the criticism encountered four years earlier when it had refused to release the results of air quality samplings following 9/11. (We now know this air was much worse than the government first claimed.) In New Orleans the floodwater definitely wasn't potable—the *E. coli* was one hundred times safe swimming levels, and lead and arsenic levels were surprisingly high—but it was not as toxic as feared in terms of petrochemicals, mainly because the service stations in the city had run out of gas before Katrina hit, and the storage tanks and chemical plants south and east of city held up well enough, with the exception of a refinery in Meraux on the eastern edge of New Orleans and two partially filled storage tanks not far from Buras, which was landfall. These tanks have the diameter of a football field; the storm surge moved them the length of a football field. (This conclusion about the water seems to contradict the televised images of people wading through an iridescent sheen of oil or gasoline. The answer, as discussed at technical length in *Environmental Science and Technology*, is that such sheens can be—and were, in most of these instances—very, very thin.

Also on Saturday, Governor Blanco hired President Clinton's highly respected FEMA chief, James Lee Witt, as her adviser, and the Department of Homeland Security issued a document titled "Highlights of the United States Government Response to the Aftermath of Hurricane Katrina."

I didn't look at it. Still haven't. They had no time to send supplies where needed, but enough time to pat themselves on the back. That very night I was trying to scrounge up insulin for diabetic evacuees in Baton Rouge! The sister of an old friend from graduate student days at LSU had flown into New Orleans from New York on the Saturday before the storm to visit her parents in Baton Rouge and ended up volunteering at the triage site at the old Kmart that doctors from nearby Earl K. Long Hospital had set up to help with the evacuees pouring out of metropolitan New Orleans. Some of these folks were in dire straits, with no food or water for days, many lacking prescriptions and other medical needs, including insulin, which had been running low and was now running out. Calls to FEMA's medical hotlines had gone unanswered, the state's Department of Health and Hospitals couldn't help, all local pharmacies had to save their small supplies for regular customers. The situation at the triage center was getting quite desperate. That's when Anne Craig thought of me. She'd seen me on some interview or another, so maybe I could help. *Maybe*, I thought when she called me. I'd sure try. I called her brother Rob, who I didn't even know was a surgeon in Baton Rouge. We exchanged quick, amazed-you're-here greetings, and he drove over to the Emergency Operations Center. After an unsuccessful encounter with a woman from the Department of Health and Hospitals who was on duty that night—she was tired, she'd been up since 5:00 A.M., she had too much work to do, we weren't the only ones with problems—Rob found the secretary of DHH, who was friendly but just couldn't help us. Meanwhile, I called my man Marc Levitan, whose wife Lilian is a researcher at the Pennington Biomedical Research Center. Did they have any insulin? Better than that, we'd hit the jackpot. Pennington had just received a large supply, and we could have it. They even delivered the medicine to Rob at 1:00 A.M., and he carried it in the family van to the triage center. Mission accomplished—by freelancers stepping in when and where the official providers had failed. That was often the case those first two weeks.

While we were waiting for the insulin, Rob told me an amazing story, also indicative of the general confusion. Two days earlier—Thursday—he had treated a nurse who had evacuated from Baptist Memorial Hospital in New Orleans. This man had just recovered the year before from a relapse of multiple sclerosis, and he had stayed till the end evacuating patients, the last one a fellow who weighed about four hundred pounds. As this guy was being lifted out of his stretcher, he fell and pinned the nurse against the helicopter. The nurse passed out from the stress, and the helicopter crew brought *him* to Rob's hospital and Rob's care. He was not seriously hurt.

The next day, while on rounds, Rob was paged by his sister-in-law, Pam, who was trying to track down an in-law who had had open heart surgery at Baptist Memorial Hospital a few days before Katrina. The last information the family had received was that there had been complications, with their in-law taken to the intensive care ward. Now they had no idea where this man was, or his condition, and they were frantic. They had tried the state and FEMA, but no one had any list of where these patients had been evacuated to. Rob immediately asked the nurse he had treated the day before if he knew anything about a heart-surgery patient who'd been in intensive care. Well, this guy did know about this patient because he had taken him by helicopter to Thibodeaux General, and he even knew the bed number in the ICU unit there. Rob phoned Thibodeaux, confirmed everything, and spread welcome good news. Other stories didn't have such a good outcome.

The following morning, Sunday, I saw the whole Katrina tragedy for myself, end to end, from the air, flying with Mashriqui, a graduate student, and the world-renowned Louisiana wildlife photographer C. C. Lockwood. We flew from Baton Rouge in the west to east of Gulfport, Mississippi. It was an almost unbelievable sight. Miles and miles of New Orleans underwater, just the roofs of the homes visible in many neighborhoods; rescue boats still hard at it, along with the helicopters; at the 17th Street breach, Chinook helicopters dropping large sandbags; the infamous Superdome with its roof half peeled back; fires burning here and there; sailboats and motorboats strewn over car parking lots; two marinas empty except for the boats on the bottom; large stretches of the Interstates underwater; the interstate bridge over the Rigolets missing numerous spans. (In an interview with Mark Schleifstein of the *Times-Picayune* immediately after Hurricane Ivan in 2004, I stated that we would have lost that bridge if Ivan had come our way. Certain state officials later said I had overstated the problem. I guess not. The bridges need to be higher and of a different design, with the road sections tied down to the legs so they cannot "float off.") The Chandeleur barrier islands were missing the northern third of their mass, with no sign of the old lighthouse my family had fished from two months earlier. There were lots of life vests scattered over the water, and I wondered if a vessel had sunk. (There were no such reports.) The beach on Cat Island had been really hammered, and the island itself had been breached on its west side, with thousands of pine trees dead from salt burn—or perhaps not, because sometimes they come back. West Ship Island was almost half its former size, and a whole lot of East Ship Island was gone, but old Fort Massachusetts was still standing.

Of course, the Mississippi coastline was devastated beyond belief for about a mile inland. The Bert Jones Marina at Gulfport, where my family had always stayed for a summer holiday, was just a pile of debris—no boats, office, restaurant, or dolphinarium (my daughters love those animals). In Mississippi it was tempting to compare what we were seeing with the photographs of the tsunami damage, but that would be very unfair, since over two hundred thousand lives were lost in Asia. Still, this scene in America was incredible, and I have no good reason to challenge the claim that it was the worst coastal damage ever in the United States.

Passing over the Lower Ninth Ward in New Orleans, I lost it. I saw some bodies in the water but did not alert anyone. I knew there was nothing we could do, and I was having a hard time controlling my emotions. All of us were, because we were not prepared. I completely understood how Aaron Broussard, president of Jefferson Parish, could have wept that same day on *Meet the Press.* "It's not just Katrina caused all these deaths in New Orleans here," Broussard told the nation. "Bureaucracy has committed murder here in the greater New Orleans area, and bureaucracy has to stand trial before Congress now." The man would not be stopped on that program. (I was in the air at the time but read the transcript.) I've cited his reports about the supplies that FEMA wouldn't let through. He also revealed that FEMA had cut the antenna wires at the parish's communication tower on Saturday. Walter Maestri, the parish's director of emergency preparedness, confirmed the episode for me. With communications suddenly down, Walter immediately called the Motorola agent, who went to the antenna site, found the parish's antenna lying on the ground and a FEMA one in its place at the top of the tower. What should we make of this? Either unbelievable incompetence or—the darker explanation I try not to believe—someone with FEMA wanted to shut down the parish's local communications. In any event, parish officials took down the FEMA antenna and remounted their own, and the sheriff posted armed guards to protect it.

Five days after the storm—beautifully warm, clear days that were perfect for the rescue effort—no one yet had a grip on things. President Bush said there were twenty-one thousand National Guard troops in Mississippi and Louisiana, four thousand active-duty military on hand, seven thousand more active duty ordered in. Maybe so, but where were they? New Orleans Homeland Security director Terry Ebbert said there were only one thousand troops in New Orleans. What a time for Mayor Nagin and Police Chief Eddie Compass to pull off what must be one of the biggest PR blunders of all time when they offered five-day vacations—even trips to Las Vegas—to police, firefighters, and city emergency workers and their families. Cops had been accused of looting, and apparently several hundred were missing from duty entirely. Now this proffered holiday in Las Vegas? It was a terrible decision from any perspective, and editorialists around the country had a ball. I don't know how many workers accepted the offer. Very few, I imagine, but it seemed like the bizarre was becoming the everyday in New Orleans, even with the city now under much tighter control. On the following Tuesday we heard on CNN that the Superdome was likely to be torn down, per a "state official." Not happening. More after-the-fact stories related more lurid tales of heinous crimes at the Superdome and the other now notorious evacuee zone, the Convention Center, both now empty. As I've discussed, many of the stories about violent crime from the first week turned out to be urban myths, and the same would hold for these new tall tales, but there was plenty of real tragedy to fill the gap.

I joined CNN's Anderson Cooper in a two-hour helicopter tour and quickly concluded that his famous editorial anger the preceding week was not a ratings pose. I was glad to see that. Oprah Winfrey went so far as to demand a general apology from the country to the citizens of the devastated region. Has that been delivered? I don't think so. At the Astrodome in Houston, where most of the evacuees from the Superdome had been transferred, Barbara Bush issued her famously unfortunate remark that "so many of the people in the arena here, you know, were underprivileged anyway, so this is working very well for them." Colin Powell said that the root problem was one of economic class, not race, with "poverty disproportionately [affecting] African-Americans in this country." Oliver Thomas, president of the city council, said that "everything the tourists want to see are still in place. . . . We have a good foundation." The story broke about the two military pilots who rescued one hundred people the day after Katrina hit, only to be reprimanded for losing focus on their logistical mission. FEMA tried to forbid reporting on the recovery of the bodies, a bizarre action by an agency that, I would have thought, would have been too busy to worry about what the media was doing. (After all, the agency had just ordered twenty-five thousand body bags.) CNN went to court, and the attempt to censor the news was set aside, of course. I took this as still more evidence that important officials with FEMA and Homeland Security were as concerned about their image as about their work.

From my outpost at the LSU hurricane centers in Baton Rouge, I looked on and listened with amazement at how badly things had gone for a nation as powerful as ours. The city was stabilized by that first weekend, but rescue efforts were still required through the *following* week. The state Department of Wildlife and Fisheries said at one point that it was rescuing 650 people an hour. That adds up to a tremendous number. The Coast Guard said it had rescued 6,000, and that number actually sounds low, because they were as organized and equipped as any agency. All together an estimated 600 boats were working the streets of the city—one number I do accept at face value. This rescue effort was noble, but it was also the straightforward part of the job, in a way: Find the folks, take them someplace dry. Almost every other aspect of the emergency disintegrated in disarray, and I became more and more determined that the truth had to get out. We had to learn from these mistakes. Right about this time, a French TV crew came and stayed at my home. I'm a nice guy, plus I'd worked with them on a warning documentary two years earlier. Now here they were again, because it had happened. They were incredulous at everything they were seeing and hearing. Of course, they were preconditioned to hold this catastrophe against America in a fundamental way. Europeans—at least the friends who e-mailed me and almost every reporter and producer I dealt with, and that's quite a few from most of the Western European countries—now look on us as strange people in a strange land. A friend in England sent an e-mail outraged by the self-serving "no one could have foreseen" protestations of Michael Chertoff. After 9/11, we had heard the same whining protest that no one could have anticipated the use of jet aircraft as weapons of terror. In fact, plenty of thoughtful people had so anticipated, just not the right ones. Or had the right ones chosen not to listen?

A week after Aaron Broussard's appearance on *Meet the Press*, it was Mayor Nagin and my turn to press the case with Tim Russert that the failure we had seen over the previous two weeks was, with the exception of first responders, systemic. Nagin said that his biggest mistake was the assumption that the cavalry would be coming immediately. The federal cavalry had always been the city's ultimate and only hope, and this hope was profoundly misplaced. "To this day, Tim," the mayor said, "no one has dropped one piece of ice in the city of New Orleans to give some people relief." I said that despite the Hurricane Pam exercise and all the warnings and the prestaging of materials, the enormity of the challenge had never quite "sunk in" with FEMA and Homeland Security and the administration. Thirteen years after Hurricane Andrew—always the benchmark for preparedness people, the ultimate wake-up call—thirteen years later and—what?

The response at all levels was still "lacking," I said, by way of understatement. "Obviously, something is wrong with the system." I proposed a cabinet-level disaster czar empowered to get the military moving immediately. (I could have added that we also need a recovery-and-rebuilding czar.) Colin Powell had been mentioned as a great choice as a disaster czar, so had Rudy Giuliani, and at some point President Bush said that Karl Rove would be coordinating everything, but in November the job would go to Donald Powell, a Texan who had served as chairman of the Federal Deposit Insurance Corporation since August 2001. I wouldn't call him a czar, however, more like some kind of coordinator and point man. Whatever his capacity, over the following months I almost never read or heard his name.

John Kerry would soon blast the failed federal response to Katrina as the rotten fruit of a "right-wing ideological experiment." On the other side of the aisle, some Republicans said it was proof that government simply isn't capable of performing at that necessary level. Trust Wal-Mart instead. Trust yourself. And yet the Republican president requested and the Republican-controlled Congress approved $52 billion in aid for the region. Such intentions several decades earlier, at all levels of government, would have saved New Orleans and cost a whole lot less. If we are ever to deal with a dirty bomb, our nation must get a grip on dealing with large-scale natural disasters. We have to learn, and learn fast.

If "learning" is actually the problem, in some quarters. We also have to care.

That edition of *Meet the Press* two weeks after Katrina may have been the last major press conference for quite a while in which the word "levee" was not spoken. Right at this point in the saga, with conditions all along the coastline stabilizing to the extent that almost everyone now had a place of some sort, somewhere, to lay his or her head at night, with the waters beginning to go down slowly as the breaches were temporarily plugged and the main pumps came on line, augmented by mobile pumps, more of us were beginning to ask the hard questions about those levees and their failure. The truly horrible fact was beginning to sink in: The entire tragedy in New Orleans—not on the coastline and in Mississippi, but in New Orleans—was primarily due to five levee failures. I have repeated this fact a couple of times already, because I've learned that many people really don't understand. Some of New Orleans was going to flood regardless, but the city could have handled it. The pumps could have drained the flooded areas quickly. Instead, we had the dreadful flooding caused by these multiple failures, and I have to admit right now that I never trusted the Army Corps of Engineers to investigate itself and find out exactly what happened and why.

Immediately after the storm, on Tuesday, I told Terry Ryder, one of the governor's senior staffers, that we could not account for the amount of flooding reported in the media, given what we knew at the time. There had to be a catastrophic breach or breaches somewhere else. Terry, whom I have known for years as a no-nonsense person, told me to go find out. We did.

In an unguarded moment Al Naomi, a project manager for the Corps, told the Newhouse News Service on the following Friday, September 9, that he assumed a "catastrophic structural failure" of the 17th Street levee. Naomi had previously been quoted in a piece for *New Orleans City Business* complaining about budget cuts of $71 million for the Corps's New Orleans district. The fact sheet on this budget cut released by the Corps in May stated explicitly that the primary remaining job in the currently authorized improvements was "the parallel protection along the London Avenue and Orleans Avenue canals. Completion of this work is scheduled by 2010." In response to the circulation of this statement after Katrina, the Corps issued a news release stating that the levee failures had been in sections where the upgrading *had been completed.* Lieutenant General Carl Strock, who runs the Corps in Washington, D.C., said the same thing to reporters: "The levee projects that failed were at full project design and were not really going to be improved." (The embankments fronting the Orleans Canal pump station—the embankments that were overtopped because they were six feet lower than the adjacent flood walls—*were* still on the Corps's to-do list.)

Looking back now, I'm surprised at both Naomi and Strock's telltale uses of the words "failure" and "failed," because it quickly became apparent in September that the Corps would argue that the storm surge from Lake Pontchartrain had overtopped the levees and scoured dangerous trenches on the backside, necessarily weakening these otherwise solid structures to the point of collapse. Naomi and Strock had temporarily diverged from the party line, which was: "The levees were sound, but the event exceeded the design. Congress told us to design to a Cat 3, and that's what we did. Our hands were tied. Katrina was a Cat 4 storm."

Simply not so, and at the LSU hurricane centers we were immediately suspicious of this whole scenario. The lowest of the levees in question were supposed to be fourteen feet above mean sea level, and the ADCIRC surge models had predicted a surge in the lake topping out at ten feet, maybe eleven, because the lake was on the western, or weaker, side of Katrina, where sustained winds were 75 mph. That's a *Cat 1* storm, folks. The catastrophic storm surge had been to the east of New Orleans and, especially, on the Mississippi coast. For New Orleans and Lake Pontchartrain and the levees, Katrina was *not* a major hurricane. It's that simple, but as I mentioned in the introduction, I know from experience that people don't want to hear this. They want to have lived through this monster, and they want the catastrophe to have been caused by this monster. In the city itself, this just wasn't the case.

In October we got word from NOAA that, indeed, Katrina might have been just a Cat 3 at landfall, and then the National Hurricane Center confirmed the number in late December. Additionally, the forward speed was fast, 14 mph to 17 mph, so over the lake and the city Katrina was a fast-moving Cat 1 hurricane—nothing approaching the storm the Corps claimed the levees were designed to withstand. (Robert Howard, an atmospheric researcher at the University of Louisiana, Monroe, believes Katrina might have been only a borderline Cat 3 storm at landfall. His analysis is complex and, at this writing, incomplete, but he might be correct. Windspeed numbers are not always hard and fast. A great deal of study is necessary to come up with the best ones, and the decisions can be changed years later. In the case of Hurricane Andrew, the official Cat 5 designation, an upgrade from Cat 4, wasn't announced until 2002, a full decade after the damage was done; Cindy was upgraded from tropical storm to Cat 1 status six months after the fact. With Katrina, we weren't surprised at the downgrading; the wind damage on the ground simply didn't reflect a major hurricane anywhere except right at the Buras landfall. Marc Levitan is a wind guy, one of the best. Touring the north shore of Lake Pontchartrain and parts of Mississippi, his team saw Cat 1 evidence—that is, roof damage to some, not even all, buildings, the odd tree down, but no catastrophic tree damage. Katrina had indeed weakened considerably in the overnight hours before landfall Monday morning.

Nevertheless, in the weeks following the storm, various voices from the Corps of Engineers kept emphasizing Cat 4 . . . Cat 4 . . . Cat 4. They wanted to fix in the public mind the myth that this very strong storm had overwhelmed the design to which they had responsibly built the levees. This myth is also reflected in an e-mail from Harley Winer, chief of the Corps's Coastal Engineering Section, to our own flood fatalities modeler, Ezra Boyd. It is dated August 2, 2005, just four weeks before Katrina. Ezra had asked whether "a major section of levee failing during a surge event is possible and . . . should be considered when looking at different disaster scenarios for New Orleans." Winer replied, "I don't think an engineered levee would fail during a storm." He added, "The federal levees are engineered and constructed to engineering standards. [T]he only levee failure is if the[y] are overtopped by a storm surge that exceeds the design." These statements pretty much sum up the Corps's attitude before Katrina and immediately afterward. Regarding the levee failures on the drainage canals, we just didn't believe it. Even at that early date, while we were still collecting data, we had a pretty good idea that our ADCIRC models had been essentially correct: no overtopping along the drainage canals in the Orleans Metro Bowl. The developers are constantly upgrading ADCIRC. They meet with a users group once a year to discuss recent developments and new ideas. Following Katrina, Jack Bevin, one of the lead forecasters at the National Hurricane Center, sent an e-mail saying that it looked like we had it right. We did, thanks to the excellence of ADCIRC and the highly accurate track and intensity predictions received from the National Hurricane Center. To get technical for a moment, one measure of model performance or accuracy is called the root-mean-square error (RMSE). If the model is perfect, with surge predictions exactly matching the actual event, the RMSE would be 0 percent. Paul Kemp has checked the Katrina models against fifty surge measurements throughout the area and came up with a RMSE of +/- 15 percent, which is extremely good. The levees on these canals were never overtopped. They failed of their own accord.

On the other hand, we couldn't really talk to the Corps directly about what we knew and what we doubted. If we had asked outright for the design and construction documents for the levees, they would never have provided them. Not then, not without pressure. Paul and I are not well-liked by some in the local district's upper management, although the more political and influential folks do like Paul. I've alluded to the combative atmosphere of the CWPPRA process in the 1990s. More pointedly, in 1993, Paul, Hassan Mashriqui, Joe Suhayda, and I had showed the Corps that they were in the wrong about the Wax Lake weir, a $36 million project that was supposed to reduce flooding in Morgan City, but did just the opposite. We got into this question because city officials and some local businesspeople asked us to. The Atchafalaya River, a major distributary of the Mississippi, of course, has two outlets at the Gulf, the Lower Atchafalaya and Wax Lake outlet. In the early nineties the Corps found itself dredging the Lower Atchafalaya River annually, because this river is the navigation route to Morgan City, an important port for shipbuilding, oil and gas, and fishing interests. The dredging cost about $4 million a year, and the Corps surmised that pushing more water down the Atchafalaya would reduce the amount of sedimentation and therefore also reduce the necessity for dredging. So it built a weir across the Wax Lake outlet, about a mile below its takeoff point from the Atchafalaya, eight miles upstream from Morgan City.

Almost immediately the local riverfront businesses, and especially the shipbuilders, were flooding more often. The Corps reminded them that it had bought the flood easements from the original landowners years before. Owning the flood easement means "we can drown you if we want to." So even if the weir was causing higher water and floods, the Corps was covered—legally. So it believed and so it said. Obviously, though, the business owners were upset that the flooding had gotten much worse immediately after the weir had been installed, and the city hired us at LSU to take up their case. We were happy to do so. The science was not all that complicated. While the Corp's idea had been to solve the sedimentation problem in the Atchafalaya by flushing more water down the channel, it was actually contributing to the problem because it had not taken into account the fact that the entire Atchafalaya Basin, which had started out as a large lake a few hundred years before, had absorbed just about all of the sediment it could and was necessarily sending the excess, in ever-increasing portions, down the river. Preweir, the river and the Wax Lake outlet were carrying sediment in proportion to their discharges; postweir, the extra volume of water in the river couldn't overcome the extra sediment.

The weir had not solved the dredging problem, but it had raised flood levels in Morgan City by about eighteen inches. With a big river eighteen inches doesn't sound like much, but to the businesses positioned right on the water, it was. We used some direct measurements and the Corps's own models to demonstrate what was happening. We also found what we thought was pretty shoddy workmanship. After an investigation and debate that lasted about a year, the Corps took out the weir at just about the cost for which it had put the thing in ten years earlier.

We had other run-ins with the Corps over the years, including a major reassessment of the levee systems in the Atchafalaya Basin. The Corps had set up a citizens' advisory panel that turned out to be a divide-and-conquer strategy, or so it seemed, because all the different stakeholders and user groups ended up at odds. Paul Kemp, the natural conciliator, helped the groups recognize the advantages of working together, and once this was accomplished the Corps was dealing with a different kettle of fish. It had to be responsive. All the while Paul, Joe Suhayda, and Mashriqui were helping the Corps with its modeling efforts for this reassessment and finding many problems, some as basic as misplaced gauging stations, major bridges, and large pilings. These made a joke of the model. Another joke: This planning effort, which has cost more than $14 million, has still not issued a final report.

Given this history it would have been very difficult for us to "go it alone" with the Army Corps of Engineers on the matter of the levee breaches during Katrina, which was the worst catastrophe to ever confront the agency.

In the weeks following Katrina, as more reporters started thinking about the official explanation that the breached levees had been designed only for a fast-moving Cat 3 storm—setting aside for the moment the fact that Katrina was not even that strong on Lake Pontchartrain—the inevitable question was, "Why protect against just a Cat 3 storm? Why leave the city of New Orleans in jeopardy against a Cat 4 or Cat 5 storm? Does the Corps think that water from the lake and storm surge from the Gulf of Mexico is somehow less wet than water from the Mississippi River? Would this water not rise as high, do as much damage, and require just as much time to pump out?"

These are not new questions, they are very important questions, and the Corps has offered an assortment of answers. Most of them revolve around an assessment-and-design process that dates from the early 1960s, an era before the serious loss of the wetlands, before modern computers, before reliable storm-surge models. At that time the levee system around New Orleans consisted of earthen embankments like those along the Mississippi River, just not nearly as large or high. None were reinforced with the steel plates called sheet piling or topped with concrete flood walls. Clearly, that system was not protective. When the Bonnet Carre Spillway from the Mississippi River into Lake Pontchartrain was opened for the first time in 1937, the lake had filled up and poured over the levee immediately into adjacent Jefferson Parish. Embarrassing. A storm in 1947 had overtopped lakeshore levees and swamped nine square miles in the city and thirty square miles in Jefferson Parish. Ominous. Sections of the levees on the Industrial Canal had failed or were overtopped (contemporary accounts differ) during Hurricane Flossy in 1956, flooding over two square miles of the Ninth Ward.

A harbinger. Hurricane Hilda had sideswiped southeast Louisiana in 1964. A reminder.

No sooner had the Corps begun thinking about doing something than Hurricane Betsy ripped ashore in September 1965, between Grand Isle and Port Fourchon (about 30 miles west of Katrina's landfall at Buras). This was a strong Cat 4 storm at landfall, and still a Cat 3 at its closest approach to New Orleans, thirty-five miles to the southwest. Even though the storm surge was only eight feet to ten feet, New Orleans East and St. Bernard Parish went under. Water from a levee breach on the Industrial Canal (there's that name again) rose twenty feet in that many minutes. Fifty-eight died, some drowning in their attics. As many as 250,000 were then evacuated. "Billion-Dollar Betsy" was the tipping point, the call to arms, and Congress almost immediately passed the Flood Control Act of 1965, ordering the Corps to build a complete system to protect New Orleans. (I've read several versions of the politics behind this measure. My favorite is the short one: Senator Russell Long, Huey's son and a very powerful man in Washington, asked his old friend, President Lyndon Johnson, to fly down to survey the scene. The president promised to send his "best man." The senator said he was not "the least bit interested in your 'best man.'" LBJ immediately called for his airplane.)

With the president's enthusiastic support Congress then ordered the Corps to protect New Orleans from "the most severe meteorological conditions considered reasonably characteristic for that region." In order to do this, the Corps commissioned the Weather Bureau (now the National Weather Service) to define the hypothetical storm that the strengthened levees would be designed to withstand and dubbed this the standard project hurricane, or SPH. In 1967, the Corps compared the SPH to Hurricane Betsy, a Cat 4, but this is just not right. The Corps now says that, taken as a whole, their defining factors amount to a Cat 3 storm—and a "fast-moving" one, which is less dangerous than a slow-moving one, as I've explained earlier.

All in all, that SPH paper tiger doesn't make a lot of sense. Some of the parameters match today's Cat 3 storm, others a Cat 4. Regardless, is this the "most severe meteorological conditions" that Congress ordered? A definite *no.* The most severe would have been a Cat 5 storm. So, the key question remains: Why was that hypothetical storm chosen by the Corps of such moderate strength and danger? Lieutenant General Carl Strock said on the Thursday after Katrina's landfall, "We certainly understand the potential impact of a Category 4 or 5 hurricane." So why didn't the Corps design the system to protect against such a storm? Why not protect one of the nation's prized cultural and economic jewels against all comers, major storms like the one that had leveled Galveston in 1900, the Florida Keys in 1935, the Texas-Louisiana border in 1957 (Audrey). Why not take the worst-case scenario and design and build to those specifications?

From one perspective, the answer seems to boil down to money. The Orleans Levee District complained that even the relatively modest design standards dictated by the SPH would be too expensive to maintain. It also opposed putting the floodgates at the mouths of the main drainage canals: too expensive to maintain. In the 1980s, the Corps stated that raising the levees around Lake Pontchartrain would be cheaper than the plan to erect barriers at the two main entrances to the lake. So the Corps may have been constrained by money, but Congress was not constrained, and Congress literally dictates the Corps's agenda. Money was not and is not the problem. Political foresight and will was the problem and will be the problem in the years to come.

Four years after Hurricane Betsy the famous Cat 5 Camille just missed Louisiana and destroyed the Mississippi coast with winds of almost 200 mph and a storm surge almost of Katrina's magnitude. A track just twenty miles west for Camille would have both leveled and drowned Greater New Orleans. No one bothered to deny it. This was in 1969. Did the Corps therefore reassess the standard project hurricane? Did it adjust its design? I wasn't here, but apparently not, even though a 1982 report by the Government Accountability Office states, "Subsequent to project authorization and based on the Weather Bureau's new data pertaining to hurricane severity, the Corps determined that the levees along the main drainage canals, which drain portions of New Orleans and empty into Lake Pontchartrain, were not high enough since they are subject to overflow by hurricane surges."

The *Times-Picayune*'s Bob Marshall drew my attention to the Weather Bureau's two redefinitions, in 1972 and 1979. For the bureau, at least, the SPH for New Orleans was now a much stronger Cat 4 storm. As the GAO report indicates, the Corps was aware of the redefinition but opted for rather minimal measures. Pinching pennies, the Corps raised and strengthened the levees at the eastern end of Plaquemines Parish. Steel-sheet piling was used to strengthen the levees along the Industrial Canal. Some of the levees along the lake were raised two feet and linked with the levees along the Mississippi River to close the western side of the westernmost bowl. On the river, the levee at the French Quarter was beefed up with a concrete revetment. The local levee boards, tired of waiting for Corps action on the drainage canals, including 17th Street and London Avenue, reinforced those levees with sheet piling, raising their effective height from plus-6.5 feet above sea level to plus-10 feet. All the work was done to the Corps's official design standards, in order to ensure that the expense could count toward the locals' cost share when the Corps finally built more substantial levees.

That 1982 GAO report was titled "Improved Planning Needed by the Corps of Engineers to Resolve Environmental, Technical and Financial Issues on the Lake Pontchartrain Hurricane Protection Project." What fascinating, disturbing reading it makes today. It states bluntly that even though the levees in southeastern Louisiana were a high priority, the Corps still had not resolved any of the named issues fifteen years after congressional legislation had mandated action; that it had completed only 50 percent of the work; that "there has been no strong effort to complete this project until recently, when preparation of design memoranda was initiated"; that completion was estimated for the year 2008. (By August 2005, this completion date had slipped to 2015.) The GAO report continued, "State and local sponsors generally agreed with our findings, conclusions and recommendations. They believe the Corps has not pursued this project with the expediency necessary to protect the New Orleans area and that only another disaster . . . would expedite project completion."

In 1985, the Corps and the Orleans Levee Board decided to solve, once and for all, the problems posed by the drainage canals in the heart of what I call the Orleans Metro Bowl. What happened next? There's a good deal of contradiction and confusion in the reporting on that subject. In any event, five years later the Corps dropped plans for floodgates at the entrances to the canals and agreed with the Orleans Levee Board to upgrade the "parallel protection plan," that is, the levees and flood walls. Now, post-Katrina, the Corps is going to spend several billion dollars doing what it considered doing twenty years ago: sealing the canals and installing pumps at the entrance structures. If the Corps and the Orleans Levee Board had moved the pumps twenty years ago, the Katrina tragedy would not have happened, because the breaches along the 17th Street and London Avenue canals were the main source of the floodwaters in the heart of New Orleans.

After a flood in 1995 (just rainfall, not a hurricane), the Corps undertook a $145 million upgrading of the pumping stations on the drainage canals in Jefferson, Orleans, and St. Tammany parishes—in theory, that is, but this initiative was never fully funded and, as mentioned, the low embankments at the Orleans Canal pump station that overtopped during Katrina were among those projects. In 1996, an attempt to study the levees along the lake broke down because of bureaucratic infighting. Joe Suhayda recalls a meeting from that era in which someone said about upgrading the levees to handle a Cat 5 storm, "We can't afford it."

Think about all that—or don't, if it makes you too angry. Then, in 1998, Hurricane Georges tacked sharply in the Gulf of Mexico, saving New Orleans yet again—and barely. Waves from the surge in Lake Pontchartrain came within a foot of overtopping some of the levees along Haynes Boulevard close to the Lakefront airport, and the surge in the lake was only seven feet. The maximum surge that struck Mississippi and Alabama was seventeen feet. This action caught the attention of Congress, which instructed the Corps to investigate providing Cat 4 and Cat 5 hurricane protection for coastal Louisiana that would encompass all of the previously authorized projects. On paper, that investigation continues to this day—although it is now mooted by Katrina, of course.

What's the deal with these studies? Why do they always take forever? I conclude that no one at the Corps took any of this seriously enough. Forget the military-management veneer. We're dealing with a classically hidebound civilian bureaucracy. Although the first storm-surge models appeared in 1979, and a highly refined and accurate one (ADCIRC) has been in hand for at least a decade, it was not until 2002 that the New Orleans district joined the modern world and began working with the latest storm-surge models. Before that, as the agency has acknowledged, it based its estimation of the adequacy of the levees on calculations that were forty years old. The Corps's Waterways Experimental Station in Vicksburg, Mississippi, did put a lot of money into developing ADCIRC in the 1990s, but the problem, as we outsiders saw it, was that the old-line engineers in New Orleans didn't have experience working in two-dimensional modeling, nor the ability to communicate with their expert counterparts in Vicksburg. The local Corps engineers complained that Vicksburg didn't answer questions, and that Vicksburg was too expensive!

Overall, the Corps of Engineers answers to whom, exactly? Congressional "earmarkers"? Lobbyists? Special interests? The demands of its own bureaucracy? All of the above and more, probably. While the New Orleans District of the Corps could not get the money to build proper levees, it got $748 million for the questionable locks in the Industrial Canal. Before Katrina the Corps estimated that a complete upgrade of the levees to Cat 5 status would cost at least $2.5 billion, perhaps $3 billion, and require twenty to thirty years. All now moot. The Corps's budget for levee work fell from $14.5 million in 2002 to $5.7 million in 2005. All now moot. Dominic Izzo, the principal deputy assistant secretary of the Corps for civil works in 2001–2002, told the *Baton Rouge Advocate* two weeks after Katrina arrived that the Office of Management and Budget in

Washington—a component of the executive branch—had targeted the Corps's funds for cuts for years, under both Democratic and Republican administrations. All now moot. Louisiana received far more Corps money than any other state—almost $2 billion, with California second, at $1.4 billion—but hundreds of millions of these dollars were for dredging and other projects that have absolutely nothing to do with flood control and not much to do with commerce. It's pork—and now moot. The total budget for the Corps is less than $5 billion annually. Compare the latest highway bill and its 6,371 earmarked projects, including the now-famous bridge to nowhere in Alaska (subsequently crossed out, although the state still got the money, as you may not have read), which came in at $284 billion. All now moot.

TEN

THE INVESTIGATION

On Thursday, September 8, Paul Kemp, Hassan Mashriqui, and I drove into flooded New Orleans to inspect the levees for the first time, escorted by an LSU cop, which was mandatory, given the roadblocks everywhere that were manned by serious military units. Our guy, Sergeant Bill Thomas, just switched on his blue police lights and the rifles waved us through. Over the following month, Bill and his lights saved us many hours. We turned off West Esplanade Avenue, and drove north on Orpheum Road on the west side of the 17th Street Canal, the dry side. For the first time we saw with our own eyes the state of the earthen embankment: pristine, I would almost say, with the grass green and the turf intact, with no drip lines or mud splatters on the concrete flood walls. As clear as day, there was all the proof we really needed that the Corps's claim was wrong. These levees had not been overtopped. From the beginning, all of us had had our doubts about the official line, but now my doubts instantly crystallized into certainty. We stopped the car and I jumped on the levee and turned to the others and said, "See, I told you we were right. Look, no sign of overtopping, none whatsoever."

These levees have three parts. The foundation is the earthen berm built on old swamp soils. Sheet piling—corrugated steel plates eighteen inches wide, fitted together tongue-and-groove fashion—is then driven into the berm to whatever depth is deemed necessary, based on a design plan that should include the geotechnical characteristics of the soil, the height of the wall, the depth of the canal, the geology, and so on. The designers decide. For many years the sheet piling installed after Hurricane Betsy in 1965 stood alone, protruding from the top of the embankment by six feet and serving these purposes: increased height for the entire levee; solid, rooted support for the berm itself; and as an underground barrier to prevent water from the canal from seeping all the way under the embankment, thereby weakening the entire structure. In the 1990s the New Orleans District of the Army Corps of Engineers affixed a concrete flood wall to the top of the steel-sheet piling—the third element. The contractor either pulled out the old steel piling and drove in a new piling or used the existing steel. In either case there is an overlap of a couple of feet between the steel and the concrete. This new structure increased the overall height of these levees to thirteen feet to fourteen feet above mean sea level. Of course, the sheet piling still served to support the whole structure, and to act as a barrier to underground water movement from the canal to the outside of the levees—and to the homes beyond. (The water level in these canals is just about sea level, fluctuating with the tides in the lake, which puts it level with the eaves of many of the homes on either side of the canal.)

I'm not a structural engineer nor an expert on levee design and construction, and I haven't spent as much time on the levees in and around New Orleans as I have in the wetlands, but I know these structures pretty well. In 2001, after our Hurricane Public Health Center was funded, I'd occasionally leave home early, drive to New Orleans, and just cruise the city, trying to feel in my veins the natural ridges, the unnatural canals and levees, what would happen during a flood. I knew about the lower section of the levee at Lakefront Airport—the one that overtopped during Katrina. I saw the weak spot at a joint between the earthen levee and the concrete wall near the Lake Pontchartrain causeway. I had seen sections of levee walls that had sagged downward, suggesting potential soil or differential subsidence issues. (Areas mostly underlain by clay will subside more than areas predominantly underlain by sand—always a concern when building levees or anything else in a swampy setting such as New Orleans.) I know that water coming over the top of a flood wall—overtopping—scours a trench in the soil at the base of the wall on the protected side, the side facing the homes. It's common sense, and it always happens.

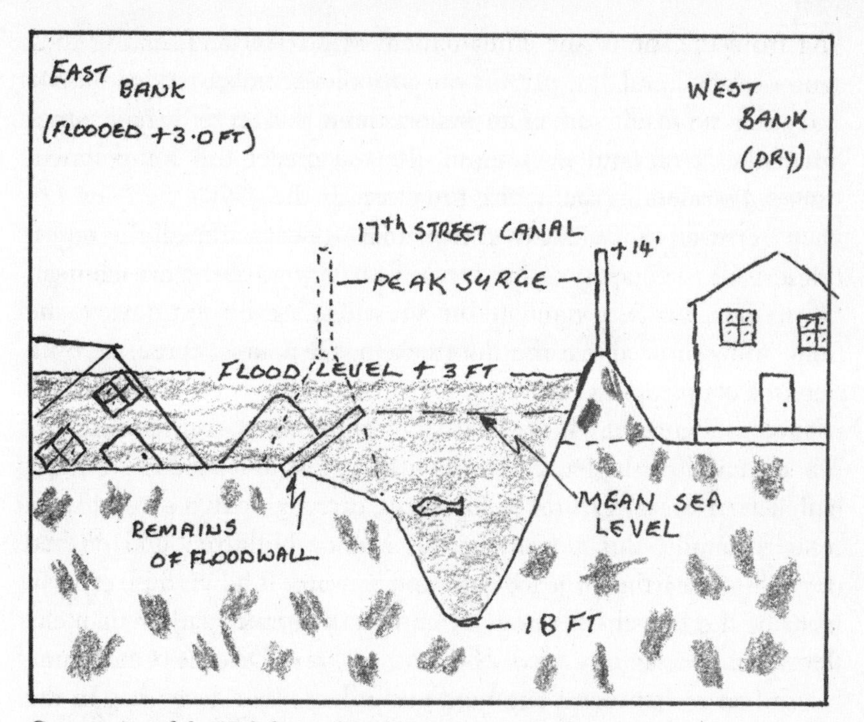

Cross section of the 17th Street Canal breach (view from the north).

There was no such scouring here at the 17th Street Canal. I couldn't believe the Corps was taking this line, because surely one of the engineers—many of the engineers—had toured the levees after Katrina and had seen the *absence* of any scouring or other signs of overtopping. Also, the new hurricane bridge just a few hundred yards inland from the lake entrance to this canal would have blocked any significant wave action in the canal itself. Waves pounding on the flood walls would not have been a factor. Nor is this canal wide enough to create significant fetch. (Wind-driven waves get bigger the greater the distance of open water. With the wind mostly blowing across them during Katrina, these narrow canals could not have supported much in the way of waves.)

What the Corps was saying seemed so far from the truth as I observed in the field that I'll admit the word "cover-up" came to my mind—and probably to my lips—as we studied the scene that Thursday. After all, the Corps and its contractors can be sued, under the

Federal Tort Claims Act. They have a lot at stake here. As the LSU group studied and compared notes, there was no question that we had all come to the same conclusion about this levee: It had somehow failed, structurally. Paul and I looked at each other and almost sighed. Here we go again with the Corps.

We crossed the bridge over the canal—the scene of all the action I described in Chapter 6—and then walked toward the breach itself on the new roadway built by the West Jeff crew, right next to the flood wall, right above the floodwaters. Of course, there were no signs of overtopping on this side of the canal either; the walls are roughly the same height on both sides of the canal. At the breach itself a section of the bank with its green grass and fence still intact had heaved laterally thirty to forty feet, carrying with it several large trees. Literally, this heave acted as a huge bulldozer and pushed everything in front of it forward and upward. The yard directly in front of the breach was now a hummocky terrain, rather than the flat, level yard so typical of this area of New Orleans. One frame house had ended up on the high ground and was fairly dry. In the scour areas on each end of the breach, houses were gone. The huge roots of the cypress trees that had grown here hundreds of years ago were scattered here and there in the scour holes. Clearly, the entire structure of the levee—flood wall, sheet piling, and earthen embankment—had succumbed to the pressure of the water in the canal and heaved laterally, almost as a unit. This breach was about five hundred feet across. Water had poured into the city with incredible power.

I met Joseph Bowles and his sister, Schoener Cole, working their way through what was left of the possessions inside Joe's flooded, ruined house near the end of the breach on the north side. This house had never flooded before, Joe said, and during Betsy this levee, just an earthen berm, with no sheet piling or concrete flood wall at all, had held, so Joe's inclination had been to ride out Katrina, just as he had stayed put for all of the other storms over the decades. Like many people, though, he had evacuated on Sunday evening, when the predictions were dire indeed. He had thought he would find a place in Baton Rouge but ended up in Arkansas, and he didn't know the fate of his house until he came back to see for himself what he had left. Even with all of the photographs telecast around the clock, he hadn't realized his house was right at the now-famous 17th Street breach. He was staying with his sister in nearby Metairie—two of the two million lives that would never be the same.

Our group now drove east to the London Avenue Canal, where there were two breaches, but with only one easily accessible, on the west side of the canal. This breach was about 450 feet across—not as wide as the one at 17th Street. The berm on either side showed no signs of overtopping or scouring. It had heaved away from the canal, shoving a little building 8 feet up and 20 feet back. (This building turned out to be the clubhouse that homeowner Gus Cantrell, a civil engineer, had built for his kids years earlier. He sent us the before-and-after pictures. Inside that house an enormous, extremely solid, and heavy china cabinet wasn't just shoved or moved by the wall of water that had roared through. It was gone.)

We couldn't know for sure yet, but at this breach, too, it certainly looked like something instantaneous and structurally catastrophic had happened on the morning of August 29, and that water from Lake Pontchartrain had poured into the city, soon meeting the water pouring in from the 17th Street breach to the west.

The third of the fatal breaks in the Orleans Metro Bowl is on the east side of the London Avenue Canal, just north of the bridge at Mirabeau Avenue. We were finally able to reach this site a week later, and we needed my Xterra to do so. Just imagine, I said to the others, I need four-wheel drive to get around New Orleans. The reason was all of the sand piled deep over an area two hundred yards or more around this breach. Dunes of sand—an amazing sight. Where had it all come from? Possibly Lake Pontchartrain. When John Pardue and I had crossed the lake on the Saturday after the storm, the echo sounder revealed that the bottom was now very irregular, with lots of dips and valleys. Scour holes? Normally the echo sounder shows a flat, smooth bottom. Since I've sailed over this bottom literally hundreds of times, I pretty much have the bathymetry stored in my head, and it appeared to me that the surge and waves of Katrina had deepened this lake—shallow to begin with, averaging twelve feet—by two feet in some areas. That missing bottom was now sediment suspended in the water, and in the canal it would have been held in suspension by the current until it spilled over the breach, where it would have deposited the heaviest material, which is the fine sand, first. That's what I was thinking as I studied the new beach inside New Orleans. But why was there so much more sand here than at the other breach on this canal or at the one at 17th Street? This was a bit puzzling, and it was puzzling because my hypothesis was dead wrong, as I learned later when we got the borehole data for this site. Below this breach the soil from minus-10 (that is, 10 feet below mean sea level) to minus-50 is thick beach sand. So this sand deep enough to cover cars was of local origin. It had come from the breach area itself. This would be a crucially important fact for the levee investigation.

Lined up with the middle of this three-hundred-foot-wide breach at Mirabeau was a slab house that had been shoved thirty or forty yards. (Or maybe it floated; entire houses can float down rivers in floods. We see it all the time.) Other slabs were now bare. Again, there were no signs of any scouring along the intact berm at both ends of the breach. No overtopping. To all immediate appearances, yet another catastrophic failure from some other cause.

A block away, Carmen and Dale Owens were working in their almost new two-story brick home. Like all of the homes throughout the city that had taken in water to eye level or higher, the first floor was a grim vision of ruined furniture and dried mud and muck on the floors and walls, juxtaposed with pictures still hanging on the walls and placed on the mantles. "Depressing" is a pitifully inadequate word for these scenes, even for an outsider who's not picking though a lifetime's possessions, looking for something to save. No wonder a lot of the evacuees now scattered far and wide were saying that they would not even come back to see their former homes. Just too difficult. Carmen said she understood this attitude, but the bedrooms and possessions upstairs were fine. They looked exactly the way they had when she and Dale had fled late Monday morning, when the winds were dying but water was suddenly in their house. They couldn't understand what had happened. This neighborhood had *never* flooded. That's why they didn't have flood insurance. They did have another house—dry—not too far way, so they had someplace to live. Still, I was amazed by their spirit. (A few weeks later one of Carmen and Dale's neighbors contacted me. They had video shot right after the breach, and it was time stamped. As our forensic studies moved forward this video would prove more and more important. It is, as far as I know, the only video taken immediately after the failure of a levee.)

Back at the breach, we looked into a house that was completely missing one exterior wall. The furnishings inside were a mud-caked shambles, of course, but there in an open closet were all the winter blankets neatly packed into plastic bags, the kids' school sports bags, Barbie dolls, everything in this one little section looking perfectly okay, but with the house around it a complete loss. My daughters loved to play with Barbie dolls, and it always amazed me how they knew which doll was whose and which clothes belonged to each one. Who and where were the little girl or girls who played with these Barbies? Did they get new ones for Christmas? The garage was filled with sand and a very strong smell of death. I said a silent prayer for this family, wherever they may be.

I've said it before, and I'll say it again now. These images stuck with me, and they convinced me to try to get the federal government to own up to the fact that this city was flooded by the failure of *its* levees. As the month passed and the unwatering of the Orleans Metro Bowl was proceeding nicely, helped in good measure by the huge portable pumps flown in by a number of European nations, more and more residents mustered the courage to return to see what they had left, if anything. My images were their homes and lives. And they were angry, too. They wanted to know what had happened. Many of these neighborhoods had never flooded in the sometimes long experience of the owners. These three breaches had effectively ruined the largest part of New Orleans for the foreseeable future. Call it a blame game if you must, but some of us were determined to find out exactly what had happened and to demand justice from the responsible parties.

The Corps had not dug these canals or built the first levees here, but when it added the concrete flood walls in the 1990s, it had conducted a comprehensive assessment of what was here and what was needed. By law and by any moral calculation you choose to perform, the Corps was now totally responsible. Whether the owners or renters of these homes had evacuated or not, they were wiped out, and I believed—and still do, of course—that the federal government needs to compensate them fully. I don't care whether I become a one-man band on this issue (but I don't believe I will). One afternoon when I was a guest on NPR's *To the Point*, Warren Olney started the program with a quote from President Bush about the need to balance compassion with fiscal responsibility. What would I like to see the president do? Olney asked me. I don't miss this kind of opportunity. Warren had really teed it up for me, and I tried to hit that ball long and straight. The feds need to step up and compensate those who trusted the levees—the federal security system—because that system had failed during Hurricane Katrina.

Like many structures, a levee under pressure is only as strong as its weakest link. In different trips to the different breaches over the following weeks and months, we looked for that link. We looked at the panels of the flood walls and the joints between. (The panels were not linked together, but stood independently with a rubber epoxy sealing the crack.) We looked at the steel rebar—that is, what we could find in the rubble—that would usually be tied with wire into a "cage," with this cage then strengthening the concrete itself and uniting the concrete in some way with the sheet piling, so that concrete, rebar, and steel piling function as one unit. We looked at the Corps's choice of an I wall for the flood wall rather than a stronger T wall with batter piles. The term "I wall" is self-explanatory: Viewed on end, it resembles the letter I, supported at its base by the vertical steel-sheet piling driven into the berm. The T wall resembles an inverted T, with a wider base than the I wall; in addition, two piles, one on each edge of the base, penetrate the soil at an angle away from the wall, almost like the legs of the capital letter A. These are the batter piles, and they assure a far more robust structure.

They are mandatory in soft, weak soils. The wide base of the T wall also helps protect the system should there be overtopping. But on these canals we have I walls.

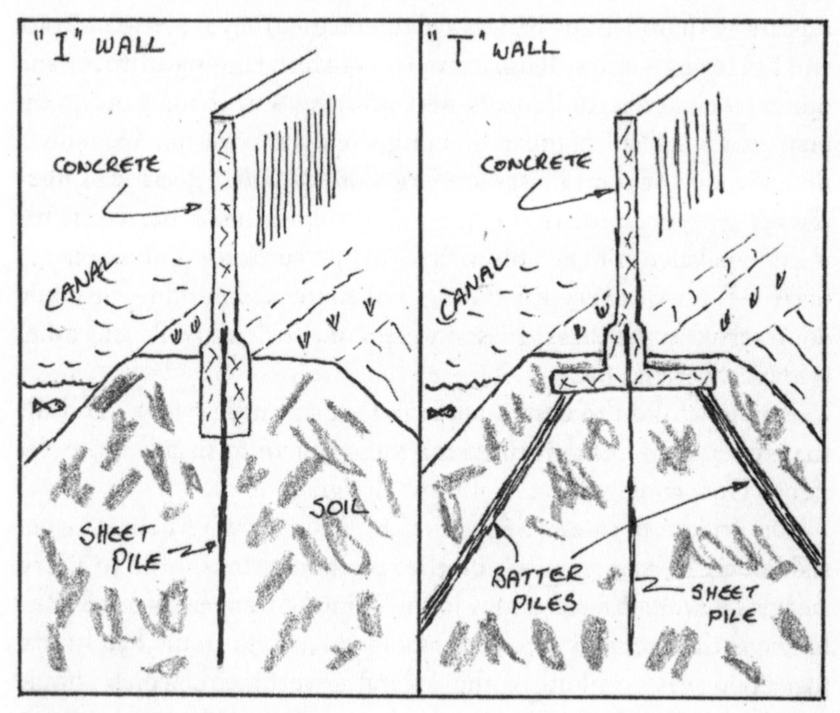

I-wall versus T-wall construction.

If an I wall begins to give way and leans back under the pressure of the water, a gap will open at its base, providing easy access for water to dig down the sheet piling—percolation, this is called. If this water gets all the way down and under the sheet piling, it may emerge on the outside of the berm as a boil. Obviously, this is a dangerous situation, with "boiling" water cutting completely through the earthen embankment. How deep in the soil was the steel-sheet piling in these embankments? Was it the same depth everywhere? Was it deeper than the canal to help ensure that no sand boils or seeps would develop under the pressure of high water levels due to the storm surge? Was any of the piling, in the failed sections, bent or distorted?

We looked at everything—or tried to, because the Corps was necessarily scrambling to rebuild the ruined berms and come up with some kind of makeshift structure that might stand a chance of holding back a serious surge, if we should get another storm, which was always a possibility. It was only September, in an extremely active hurricane season. Barges and trucks were hauling in rocks and other fill material, bulldozers were shoving and shaping new embankments, and all of this rebuilding work was covering up some of the evidence of the failures. Sections of the failed flood wall were buried. The structure of the original heave was being buried. What if the equivalent of the "black box" or the smoking gun was being buried? This was very frustrating, but there was nothing we could do. Every day that passed less and less of the flood walls and other evidence were visible.

And what did the design plans and specifications show? Did the structures, as we could still piece them together, match those designs? How could we ever get those designs?

On Friday afternoon, September 16, I talked with Michael Grunwald of the *Washington Post*, on the recommendation of John Barry, author of *Rising Tide*. I'd met John during our appearance on *Meet the Press* the previous weekend, and we'd agreed immediately that this flood was a failure of the federal government, which should compensate homeowners and businesses for their losses. Barry would soon publish an Op-ed to this effect in the *New York Times*. But now I needed a national reporter who was following the Katrina story and would be open to our tentative interpretation of why the levees had failed. In a few TV appearances I had tried to get the debate going, but no one seemed to realize its significance. I wasn't getting a lot of traction, as they say. The overtopping explanation from the Corps was misleading and self-serving, because it allowed them to claim that the surge exceeded the design capacity, which would let them off the hook. (Who knows, they may still try to argue this, but without a shred of evidence, it won't be easy.) I was concerned that good forensic evidence would be buried at the breach sites or otherwise lost before any congressional investigation could gear up. I thought we had to get the alternative story out there immediately for Congress and the world to consider. What better way than through the *Washington Post*? I hated not bringing in the *Times-Picayune*, specifically Mark Schleifstein, who is a great comrade-in-arms in the hurricane business, but I thought, rightly or wrongly, that the *Washington Post* would be the best mouthpiece. I wanted everyone in Washington, D.C., to see what was going on and above the fold, if at all possible.

John Barry agreed with me and set up a conversation with Michael Grunwald. On the phone Grunwald was very friendly and polite, heard me out, and seemed to understand right away what I was saying and the gravity of the situation. He said he needed to talk to one of his editors, but would try to fly down as soon as possible. Apparently his editor got the picture too, because I picked up Grunwald at the Baton Rouge airport on Monday just after noon, then collected Paul Kemp and Mashriqui, and we drove straight to New Orleans and toured the 17th Street and London Avenue breaches that afternoon. Level-headed Paul Kemp said straight out at one point, "This should not have been a big deal for these flood walls. It should have been a modest challenge. There's no way this storm should have exceeded the capacity." That was what we wanted to demonstrate to our guest from Washington. We returned to Baton Rouge, had a great dinner on Grunwald's newspaper, then set off early Tuesday morning for the Industrial Canal. That afternoon Grunwald called spokesman Paul Johnston with the Army Corps of Engineers one last time to get comments on the alternative theory shaping up with the LSU hurricane experts—and others—that overtopping had had nothing to do with the three failures on the 17th Street and London Avenue canals. I sat there listening to one end of the conversation, during which Johnston repeated yet again that "the event exceeded the design."

Grunwald talked to former Louisiana congressman Bob Livingston, a Republican who had chaired the House Appropriations Committee and was now a lobbyist in Washington. Livingston, who must understand well the symbiotic relation between the Corps and Congress, noted that the earthen levees along the lake had held while these smaller berms topped by flood walls along the canals had failed. "I don't know if it's bad construction or bad design, but whoever the contractor is has a problem," Livingston said. So does the Corps, I thought, when I read that quote in Grunwald's story, cowritten with Susan Glasser. It ran on Wednesday, September 21, with the headline FAULTY LEVEES CAUSED NEW ORLEANS FLOOD, EXPERTS SAY. And yes, it was above the fold! Since this was my birthday as well, I took the story and its location as a good omen. The following day Mark Schleifstein laid out essentially the same contrarian point of view on the *Times-Picayune*'s Nola Web site. In that story Colonel Richard Wagenaar acknowledged that the overtopping theory was being questioned. He said that the Corps had been ordered by Congress to retain all documents. He added, "My guys want to know what caused this just as bad as everyone else, because we've got two hundred people on the other side of that canal who lost their homes as well—at least two hundred people." He reiterated, however, that the design criteria may have been exceeded by this storm. He pointed to debris inside the lakefront levees at Metairie, in Jefferson Parish west of New Orleans, proving that waves had reached at least seventeen feet in that section of the lake.

Not really. We had also looked at that debris, and it told us that the waves were dumping debris on the lake side of the earthen levees, with winds of up to 70 mph lifting this debris and blowing it over the levee. There was no evidence of overtopping here, as perhaps the colonel was suggesting.

Without a doubt, I think, the batch of overlapping stories in the press turned the corner for us and guaranteed that the truth about the levees would come out, sooner or later. Calls and e-mails from reporters picked up, that's for sure. My phone rang at 2:00 A.M. on the morning the *Post* story appeared. I was under the weather and didn't answer, so they called Paul Kemp. Being the tough soldier he is, Paul left his home in Baton Rouge at 4:00 A.M. for a dawn interview in New Orleans with ABC—all part of a great team effort from day one until the present. Our angle about the levees and the Corps gained even more momentum when, unbelievably, sections of the St. Bernard Bowl and the Orleans Metro Bowl, both almost drained from Katrina's floodwaters, took heavy water again on Friday, September 23, as powerful Hurricane Rita, passing 150 miles to the south on her way to a landfall in far western Louisiana, threw up a seven-foot surge, and the Corps's breach repairs along the Industrial Canal proved unsatisfactory.

Rita was scary. When she first shaped in the Caribbean the previous weekend, the most cursory consideration of the weather maps put a lump in the throat. This storm, too, gave every indication of floating west, just as Katrina had, right into the Gulf of Mexico for the almost predictable sudden strengthening, and then at some point starting her swing to the north. Of course, any kind of reprise of Katrina's path—or, even worse, one somewhat to the west—might have been the bitter end for the city of New Orleans. The levees are "severely degraded," I told one inquiring reporter, and the city is "extremely vulnerable." To put it mildly. On the Industrial Canal, the quickly repaired levees on the east side were several feet lower than the originals. The Corps said these barriers, composed of rock and limestone chips, would be high enough for the predicted storm surge. They weren't, and some sections eroded as well, with water pouring once again into Lower Ninth, Arabi, and Chalmette. (We had warned the Corps that the limestone chips should have been protected with surface seals.) The smaller breaches on the west side of Industrial Canal were also overtopped again, and water returned to the adjacent neighborhoods, including Gentilly, in the Orleans Metro Bowl. Fortunately, the 17th Street and London Avenue canals had by then been sealed near the lake entrances with sheet piling driven down alongside the bridges—a makeshift mechanism whose function should have been in place all along and that must be a feature, in one form or another, of the upgraded levee system everyone is promising will now be built.

With the levee investigation events followed hard and fast for the next six weeks or so, as everyone came to appreciate how important the answers were. Three groups were officially investigating what had happened: one team sent in by American Society of Civil Engineers (ASCE); another from the University of California at Berkeley, working under the auspices of the National Science Foundation (NSF); and, as of Wednesday, September 28, the Corps's official investigators (the Interagency Performance Evaluation Task Force), with its report not due until June 2006! A fourth team was our group from LSU, but we were unofficial. I didn't like this. Without an official mandate to do a forensic study, I feared hassles from LSU, where some people apparently were getting nervous about the involvement of us "hurricane center people" in the levee failure story. In fact, just two doors down from my office sit some folks who try to get money out of alumni, some of whom have made huge fortunes in the local petroleum and chemical and shipping industries, and word filtered out that some of these alumni were upset about our visibility. My answer was, tough, but I also knew we needed some kind of independent status, so I worked the phones with state people, pointing out that the ASCE and Cal Berkeley teams didn't have their roots in the details of Louisiana geology. And were these people truly independent? After all, former Corps employees were among the members of the ASCE team. What credibility would such an investigation have? I thought the state of Louisiana needed its own official investigation, beholden only to itself, and my main sounding board for this was Terry Ryder, someone I have known for a number of years. Separately, Johnny Bradberry, the head of the Department of Transport and Development (DOTD), had been moving on a track he called "Team Louisiana." I was not surprised. Secretary Bradberry is extremely sharp and knows how to make a decision (so important in this time of crisis), which I guess he learned from years of managerial experience in the oil patch. I knew we'd be able to work together.

As noted, our LSU team has never enjoyed smooth sailing with the Corps, at least not for long. I didn't help our cause when I suggested at a getting-to-know-you dinner with a group of Corps officials at a Baton Rouge steak house that the site of the levee breaches should be treated as a "crime scene." Typical bluntness and perhaps an unfortunate choice of words. All I wanted to convey was the need to look at all the evidence at every breach before still more of it was covered up by the ongoing repair work. We needed to get everything to an off-site building, where all of us could have easy access, including, eventually, the media and the public. Why did the Corps choose that meeting to offer us the use of a valuable centrifuge? Were they trying to get on our good side? Paul Kemp said after the meal, "Well, all we learned tonight was that the Corps likes to eat steak." That cracked me up. We were supposed to meet these guys in the field the next day, but they were no-shows. It was so strange with these engineers: From one day to the next we never knew what we were going to get.

Then someone with the state was told that someone with the Associated Press said that I had called the Corps "nonprofessional." That certainly would have been my thought, but I didn't say that in the interview in question. The reporter might have asked, regarding some action or another (he would have had numerous options), "Isn't that unprofessional?" I might have answered "Yes." Live and learn, but that's one reporter I won't talk to anymore. The damage was done, however, and a state official whose support we needed was upset, so I sent around an e-mail apologizing for any misstatement on my part. (I also sent around e-mails suggesting that the state set up a new coastal restoration czar—a real one, this time, with real powers.) Thank goodness none of this sabotaged the Team Louisiana idea, which was soon ready to go. Before the official announcement I sent Paul Mlakar, head of the Corps's investigative team, an e-mail advising him of the new group and urging full cooperation and sharing of data. Mlakar replied that the Corps had made contact with LSU "last month" about a "joint data-collection effort and have included LSU most intimately in ours thereafter. . . . However, not all in the engineering profession believe that the reason for the disaster is as clear as you suggest." Then, out of the blue, on Saturday morning October 1, we got a call informing us that the Corps was going to begin a survey of high-water marks in two hours. Did we want to join them? How strange: two hours' notice on a Saturday morning, one that also happened to be the first day off most of us had had since the week before Katrina. What we didn't know was that the ASCE and Cal Berkley teams were in town for the first time to kick off their own investigations, hosted by the Corps.

Paul and Mashriqui immediately dropped family plans and rushed down to New Orleans. Paul called me that evening with the news that the Corps had organized a big meeting of all the teams for the next evening, Sunday. I wanted to be there—and I should have been, since I'd be heading the new Team Louisiana. But the damnedest thing happened. I was told there wouldn't be enough room for me at this large meeting of Corps officials and the investigative groups. There was room for Paul Kemp, but not for me? The meeting was at the Corps's district office, where there are some huge assembly rooms. Subsequently, members of both of the other teams confirmed that the room had been crowded, but that there had been space. Of course there'd been space! This was so childish of the Corps. I think the other teams were appalled at what had happened, and they told me that some of the Corps's people had explicitly said that they hoped these other teams wouldn't be like me—that is, full of bothersome questions.

On Tuesday, October 4, we met with staffers for the Senate Homeland Security Committee who were in town for some preliminary fact-finding prior to the committee's upcoming hearings about the federal government's response to Katrina. We put together a CD and a movie that summarized where things stood, in our estimation, and then I was very surprised—more than that, distressed—when, out of the blue, I thought, Paul Kemp brought up the idea that "harmonics" in the canal waters might have caused, or at least contributed to, the collapse of the levees. The Republicans in the room seemed to perk up at this possibility, and I sensed an attempt to get the Corps and the federal government off the hook. Later that day Paul explained that he was just trying to be open-minded because we had not completed our field investigations. (The harmonics argument went nowhere immediately.) We went to dinner with some of the staffers and tried to figure out the political dynamics. My initial conclusion was that there was a split along partisan lines, but, in the end, I decided that the whole group was rather nonpartisan, commendably so.

The following day we drove these folks to New Orleans, and I had a chance to discuss some of the many issues about Katrina that from the perspective of a disaster science specialist were glaring examples of how not to run a response operation. I told them about the ridiculous episode of the previous weekend, when I'd been excluded from the Corps's meeting. Touring the breaches and the drowned city, the staffers were suitably stunned, even though the buildings didn't look to me as dark as they had just a few weeks before. Still, it was unspeakably desolate, with piles of trees and limbs and garbage everywhere (twenty-two million tons of garbage, by one estimate, easily the biggest cleanup job in American history, including the Twin Towers); boats still parked in the unlikeliest spots; mounds of discarded possessions on every overpass; block after block after block with no people whatsoever; street corners festooned with signs advertising cleanup or restoration work (imagine the scams), and lawyers. At one house someone had placed a Santa Claus by the front door. Was that for Christmas '05, or '06? or '07?

Coincidentally, we ran into both the ASCE and the Cal Berkeley investigators at the London Avenue breach. I thought they were none too friendly to us initially, but I was told later that others in our group had good exchanges with them. Paul Mlakar was also there. The Corps was preparing a handbook for everyone, he said. I took the opportunity to ask him, in front of the folks from Washington, about the Sunday meeting to which I hadn't been invited for space reasons. "It was tight in there," Mlakar replied, but he then assured me that LSU would be invited to the forthcoming meetings.

I guess it's clear that the "banning" episode had really soured me. I decided things weren't quite right in New Orleans, and the first chance I got that day I told Marc Levitan about my concerns that the Corps might interfere in the investigative process, that they were exerting an overbearing presence and attitude, and, most important, that there might be interference from some members of the ASCE headquarters staff. I told Marc that if I called Michael Grunwald of the *Washington Post*, say, with the story about what had been going on with the Corps, topped off by the refusal to allow me in the big meeting, that story would be news two days later. If the ASCE were ever seen as a tool of the Corps, it would be putting itself in a position to get hammered. Marc's face dropped. He is well connected with the society, and he knew that I wasn't bluffing, and he knew my analysis was right. Shortly thereafter, Paul Mlakar was a lot friendlier to me.

But get this: On that very day in New Orleans our geotechnical engineer, Radhey Sharma, was invited to join a working session that evening of the ASCE investigative team. Radhey is an ASCE member. In the afternoon he broke away from us and drove away with some of those engineers. A couple of hours later, he called me. Standing right in the doorway of the meeting someone from the ASCE team had rescinded Radhey's invitation, because of "things difficult to make public at this stage." So we doubled back to pick up Radhey on the steps of the building in which the meeting was going on. I think he was suddenly unwelcome because the ASCE team was going to put together a press release for the next day via a conference call with ASCE management in Washington. They didn't want anyone from Team Louisiana hearing that conversation (which lasted for two hours, I understand).

On Friday I was tied up with a crew from the PBS series *Nova*, but Radhey, Mashriqui, and Paul headed down to New Orleans very early for a meeting prior to the ASCE press conference, where a senior Corps official introduced everyone except the LSU team members. Paul Kemp abruptly stood up and pointed out that not everyone in the room knew the group from LSU, and he then asked Mashriqui and Radhey to introduce themselves. My man.

I was working with *Nova* when Paul called with the news that the ASCE would announce that the levees at the London Avenue and 17th Street canals had *not* been overtopped. Vindication for us, that's all there was to it. That evening Mashriqui hosted two of the ASCE members at his home for a barbeque, one from Japan, the other Professor Jurjen Battjes, one of the world's leading levee experts, from the Netherlands. Now we learned that the official statement from the ASCE team had indeed been watered down after the phone call with the front office. The phrase "soil failure" had been deleted from the statement. I listened to everything they had to say, and I came away with the clear impression that at least some of the field engineers were extremely unhappy with the way ASCE was dealing with them, their findings, and the whole situation. A few weeks later one of the ASCE team e-mailed me with the remarkable revelation that all eleven investigators had threatened to resign, standing out there on the levee, unless the interference in the process was stopped. Jurjen Battjes told us how he had disagreed with a further watering down of the statement about the nature of the failure. The wall slid. It failed. That was right in front of everyone's eyes. Finally, Jurjen was simply appalled at the design of the levee systems in New Orleans.

Within just a day or two of the release of the watered-down ASCE statement, Team Louisiana was officially in business under a contract with the Department of Transportation and Development. We were the LSU guys and three private-sector civil/geotechnical engineers with truly vast experience with the local geology—Billy Prochaska, Art Theis, and Louis Capozzoli, whom the rest of us soon called the three wise men. (The contract also included small amounts for surveys, for our stopped-clock program, and for a program to capture the oral history of the survivors—these last two part of the effort to pin down as closely as possible the exact time for each failure.)

Now there were *four* investigating teams in the field—ASCE, Cal Berkeley (NSF), the Corps, and us representing the state—and now everyone would be able to collect all the forensic data. We'd have access to the original design plans and specs for the levees. Or would we? When Paul Kemp soon asked for some surveying data from the Corps that he had helped collect, he was advised to submit a FOIA request—under the Freedom of Information Act. This, Paul replied, is "inappropriate." I was amazed by his self-restraint. I might have said, "Look, asshole. Give us what we need, what you've been saying all along you'd give us." Paul took the milder approach, probably wisely, and received an immediate acknowledgement that the FOIA answer had been wrong. The survey data would be forthcoming. Over two weeks later Paul had to send the Corps another note: Now, where's that surveying data? Meanwhile, we read in the newspapers that the Corps would be making available to the investigators (or some of the investigators?) 235 boxes of material. Some of these documents have been posted to the Web, and some of these have key pages missing, such as calculations on lateral stability at the 17th Street breach.

Even after we had official status as Team Louisiana, we had to fight for every scrap of information. We even had to rely on the media. It's almost embarrassing to admit this, but so it was. Sometimes we would point a reporter in a certain direction, at other times a reporter or television producer would find new data, bring it to us, we would comment, and then they'd release it to us. This is how we were forced to build up most of our background information. Then we found some records in the New Orleans office of the DOTD. Getting data out of the Corps was extremely frustrating. We never seemed to get what we asked for, but we were hearing through the grapevine that the Corps's own "independent" investigators had access to boxes and boxes of stuff.

Nor were we alone in battling for the necessary data. The Berkeley levee warriors, as we soon began calling them, were having their own problems with access to the Corps's documents, even though they were working under the auspices of the National Science Foundation, and even though Bob Bea on that team had started his engineering career with the Corps in south Florida in 1954, building levees, canals, and pump stations. His father was a career Corps employee, so Bob truly understands the unique bureaucratic psyche that confronted us. He also has a visceral connection with the Katrina tragedy, because he lost his home in New Orleans when Betsy flooded parts of the city in 1965. He had waded to his house in the Lower Ninth to salvage what he could, which was nothing, not even his wedding photographs. He sold that house for the price of the land—a seriously discounted price.

I've mentioned the press conference for the release of the preliminary ASCE report on the levee failures. The night before, one of the people helping to set up the event had collared Bob Bea and pointedly asked if he was going to attend. Yes, Bob said. Perhaps he shouldn't, this person said—more than once. That's not the way to deal with Bob Bea. He was there the next day—and so was his antagonist, standing right next to him. Was someone hoping to muzzle this man? Forget it. He soon met with reporters from the *Los Angeles Times* and elsewhere, and all the coverage over the weekend stated quite clearly that there had been a massive failure of the levee flood walls. Bob was joined in this PR campaign for the truth by Ray Seed, another member of the Berkeley team whose father, Harry, was one of the great geotechnical engineers, now often considered the father of geotechnical earthquake engineering. (Harry Seed was awarded the National Medal of Science by President Reagan.) As Ray tells the story, he was fiercely determined to be anything but another "geotech," but now he occupies his father's old chair at Berkeley!

And people thought they could manipulate the investigation and the judgment of Ray Seed and Bob Bea? Instead, these two scientists helped us carry the banner during the fall, supporting our earlier conclusions, helping the world realize that overtopping did not cause the tragic breaches along the 17th Street and London Avenue canals.

Believe it or not, this is how things went with the investigation of one of the most catastrophic levee breaches and floods in American history—an investigation that was not just an academic exercise, because more storms are coming, and that same levee system is still the only one we have.

In order to set the stage for the next phase of the investigation story I need to backtrack briefly. On September 30, NBC News had broken the story about an old lawsuit against the Army Corps of Engineers regarding work that dated from 1993, when the levees were upgraded. In this action the Pittman Construction Company complained that the Corps had not provided enough information about the problem with the soil at the 17th Street Canal, which turned out to be weak and shifting and "not of sufficient strength, rigidity, and stability." These poor soils caused problems pouring the concrete flood walls, which was Pittman's job. Also, the firm alleged problems with the steel-sheet piling. The dispute concerned just 12 of 257 flood-wall sections, those which had tilted a little and were technically "out of tolerance." The Corps had allowed a variance on the

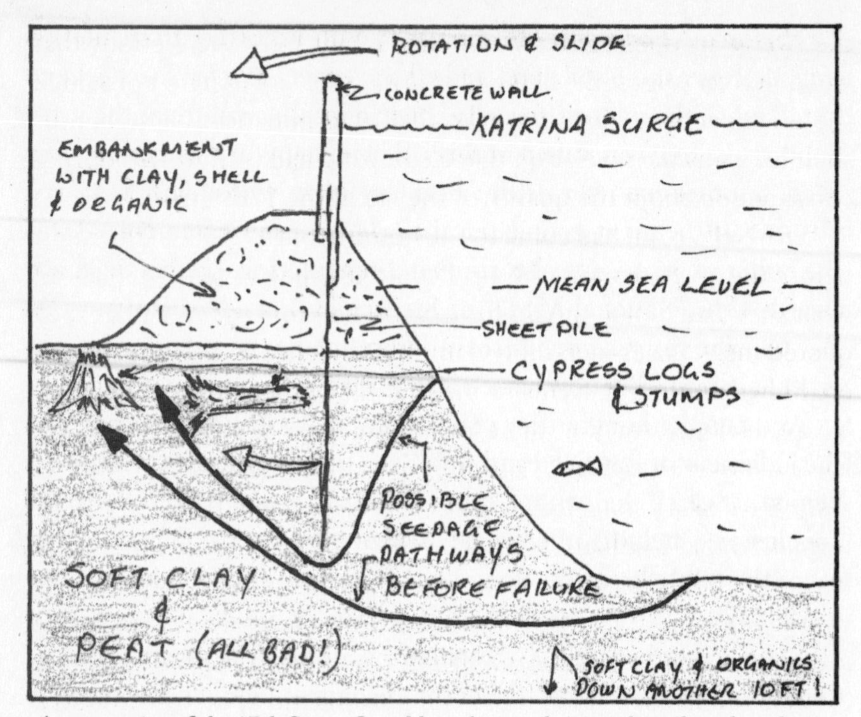

A cross section of the 17th Street Canal breach area showing the soils and predominant failure mode.

work and accepted the job, but the company claimed that it had incurred huge cost overruns and was seeking relief from the Corps.

I had not heard of this lawsuit. None of us at LSU had. At this point in the investigation of the breaches the Corps had not released any design or construction documents (and it would not for another month). As I've said, we were forced to get much of our information with the help of the media, who were now out in force on this story and digging furiously, as this NBC scoop made clear. I also suspected that the Corps was getting "leaky," that at least one of the two hundred or so Corps employees who had lost their homes was divulging information.

These courtroom documents had been filed in 1994, and the administrative law judge had ruled in 1998—in favor of the Corps. For the purposes of the investigation of the Katrina flooding seven years later, the decision itself was irrelevant. Almost irrelevant was whether the twelve sections addressed by the lawsuit were the sections that had failed on August 29, 2005. (Naturally, the Corps hasn't provided the documents we need to make that determination.) The bombshell here was the revelation that a company had complained during construction about the quality of the *soil* in the 17th Street levee.

The NBC team had come to my home about 11:00 P.M. the night before the piece went on the air. I could tell they were excited about what they had. They should have been. As I told them, and as they quoted me, "That's incredibly damning evidence. I mean, really, incredibly damning. It seems to me that the authorities really should have questioned whether these walls were safe." Bob Bea, the Berkeley levee warrior, agreed. "I think it is very significant," he said. "It begins to explain some things that I couldn't explain based on the information that I've had." Indeed, because the soil that comprises the berm that holds the sheet piling that supports the flood wall is, obviously, all-important. If the weakest link of any structure is the foundation, that structure is a disaster waiting to happen, and that disaster will happen at the most inopportune moment, by definition: in the case of a levee, when it is under pressure from high water.

The judge in the Pittman case may have found for the Corps, but all of us investigating the Katrina flooding sensed a breakthrough here. Recall that this entire city was once a cypress swamp, as indicated by the stumps and logs exposed by the upheavals at the breaches. What happens to sheet piling driven into one of these old logs, which are spaced probably every fifteen feet or so? The steel sections could bend or split apart at the tongue-and-groove sections, and there goes the barrier against seepage. What about the soil itself? The Corps's original environmental impact statement for the levees, dated 1982, describes the soil in the lakefront area as "a thin layer of very soft clay underlain by silty sands" and "peat and very soft highly organic fat clays." The original soil material for these levees was this same dredge material. Peat is organic matter. It decomposes, compresses, subsides (the New Orleans problem in a nutshell). It is an excellent path of least resistance. Under pressure such as that in a storm surge, water moves very easily along layers of peat. It can create a slippery surface. Structures built on such soil can tilt, break, slide, float, sink. Peat must therefore be carefully considered before construction, and engineers must look at the "global stability" of the whole. What is the history of the compaction of these berms since they first took shape in the first half of the twentieth century? How closely did the Corps test these soils before the flood-wall upgrading program was initiated in the 1990s? Did it account for the peat in the calculations? Did it account for the effect of adding the heavy concrete flood wall to the sheet piling? Did they pay heed to Engineer Manual No. 1110-2-2502, which has language that basically states that if you build a flood wall to fourteen feet, then it must hold water up to fourteen feet without failure? Did the Corps pay heed to the design memorandum for the MR-GO levees that state those levees should be built to withstand overtopping such that overtopping will not endanger the security of the structures or cause material interior flooding?

In the last chapter I mentioned the 1982 GAO report titled "Improved Planning Needed by the Corps of Engineers to Resolve Environmental, Technical and Financial Issues on the Lake Pontchartrain Hurricane Protection Project." I had found this document late one night while searching the Web, terminally frustrated with the Corps for not releasing its records to us. One of the most damning elements of that report, in retrospect, is the Corps's attribution of schedule delays "primarily to unforeseen foundation problems. . . . Increased construction time for flood walls, levees, and roads as a result of foundation problems discovered after project initiation." There it is, in black and white. The Corps of Engineers was well aware of foundation problems years before it constructed the flood walls along the London Avenue and 17th Street canals. Why, then, did it change some of the designs to call for less robust structures after this date? On Tuesday, October 11, Ray Seed from Cal Berkley fired off a letter to the Corps, pointing to unstable soils as the source of 17th Street and London Avenue failures. He linked the evidence to the Pittman lawsuit.

As the soils in the embankments came under suspicion during the investigation, so did the depth of the sheet piling, with the first point being a very simple one: The steel must be driven below the softest of the soft stuff. Clearly, it should not be *rooted* in the soft clays, peats, and muck that will just not support the whole structure when high water in the canal adds side forces to the equation. Drive the piles until you hit solid earth! If this is minus-80 feet, then so be it! Spring for the extra steel! The very life of a city hangs in the balance. You don't need to be a structural engineer to understand all this.

The second point is also simple: Even if the soils were the best possible, which they definitely were not, the sheet piling should be at least as deep as the bottom of the adjacent canal—minus-19 feet. What was the case with these levees? We eventually determined that the original design memorandum for the 17th Street Canal called for sheet piling driven only to minus-10 feet. Not nearly deep enough. Reporters were telling me that Corps officials were saying that the steel had been driven to minus-17, so I asked them to get the Corps to supply the records, called Pile Push Lengths. As of early February 2006, we still had not seen the numbers, although certain Corps staffers working at the breaches told us that the district office had the data. Of course it did.

Because pulling the sheet piling to determine its depth would destroy the flood wall, we needed a nondestructive test, and still somewhat experimental sonar testing seemed the way to go. We'd also use a cone penetrometer setup to get data on the strength and other characteristics of the soils. We got the funding for part of our investigation in three hours, working with Secretary Bradberry at the DOTD and the state attorney general's office.

For the sonar test, the investigators, Southern Earth Sciences, pushed a probe equipped with a special sound recorder—a transducer—into the soil, no farther than six feet from the wall. The basic idea of such testing is that at each one-foot increment of depth the concrete flood wall is struck with a special hammer, and the transducer in the ground records the signal. As the recorder is driven deeper and deeper, the signal doesn't change much *until* the receiver is lower than the sheet piling. This change should tell the investigators the depth of the steel at that location. We knew that the Corps was conducting similar tests using a different system, and we believed ours was superior, because, for starters, our recorder was quite close to the wall, not as far away as twenty-two feet, as with the Corps's system. However, both groups were going to rely on the same Colorado laboratory, Olsen Engineering, to interpret its data.

That lab told both us and the Corps that the steel ended at minus-10 feet—eight feet shy of the bottom of the canal, and not nearly deep enough. Both we and the Corps reported this number to the media, but probably thanks to the Pile Push Length numbers that the Corps has but have not released to us, it had reason to believe that the engineering firm had misinterpreted the data. The sheet piling was at minus-17, the Corps believed, and in December it orchestrated a widely covered media event at which sections of the sheet piling were yanked dramatically from the levee by a large crane. I was in The Netherlands at the time, studying their excellent flood-control system, but I heard and read later about the joy on the faces of Corps officials. This particular piling had been driven to minus-17. The Corps then said our data was bad. After careful analysis of our methods, we thought the data must have been misinterpreted by the engineering firm, and after a reevaluation, owner Larry Olsen realized that he had indeed misinterpreted our data. Subsequently, he flew down to New Orleans and went out in the field for two days with us and Scott Slaughter of Southern Earth Sciences, all of us hoping to better understand how this embarrassing error had occurred. Also along on those two days were investigators from the state attorney general's office, who regularly accompanied us in the field, in case there was any subsequent ligitation between the state and the federal government. They wanted to observe the retesting of the sonar technique.

We used different techniques to collect a series of sonar data sets. In the end, after some discussions in the field, Scott Slaughter came up with his own methodology for the seismic interpretation, which we believe is pretty much foolproof. Really, though, whether the steel in a given section of the canal levees is minus-10 or minus-17 is almost moot, because even minus-17 is not deep enough, not when the channel is more than eighteen feet below sea level and the soils are mucky clays and layers of peat. As Billy Prochaska, an absolute geotechnical engineering whiz, stated clearly in one of his reports, strength calculations show that the flood walls would fail when the water level in the canal approached 11 feet (the Katrina surge, as it turns out), no matter if the sheet pile was minus-10 or minus-17. Billy believes the Corps's engineers failed to consider the weak soils between minus-10 and minus-25 feet when they did their strength and safety calculations. Later, the Berkeley levee warriors came up with exactly the same results.

Modjeski and Masters, the engineering firm that developed the design for the flood wall on the 17th Street Canal, said in a letter to the *Times-Picayune* that it had recommended that the sheet piling be driven to a depth of minus-35 feet, but the Corps requested a shallower depth. The Corps responded to the paper with a written statement, defending its own actions. One point is certain: As part of its latest temporary patches at the breaches, the Corps has driven the new sheet piling to a depth of at least minus-50, sometimes minus-65.

Then in early February 2006 we learned from documents provided the Cal-Berkeley team that a cross section in the design documents for these levees shows the peaty soils between minus-11 and minus-16 feet, while the soil borings on which the cross section was based found the peat as deep as *30 feet.* Excellent detective work by J. David Rogers pointed out this geotechnical data irregularity. Our engineers agree with the Cal-Berkeley engineers that the Corps's levee designers may have designated minus-17 for the depth of the steel-sheet piling because the erroneous cross section showed that to be the bottom of the peat layer. Was this breach at the 17th Street Canal caused by a simple error in transferring the raw data to the visual depiction in the cross section? Of this we can't be absolutely certain, but the error didn't help.

Every day that fall and early winter, new data was announced or old data uncovered by one source or another. Every day, new statements and new stories hit the press, and with almost every one the Corps's valued reputation for conservative overdesign took a serious hit. For all of the criticism that agency has endured over the years, cutting corners on design has not been one of them, for the most part. The rationale for a project may have been dubious, or the project may not have worked as advertised (the Wax Lake weir described earlier is an example), but rarely has the design been on the cheap. At least, that's been the reputation until now, when *any* reasonably objective observer would conclude that the soil structures for these vital canal levees weren't strong enough, period, and the sheet piling wasn't driven deep enough, period. Terrible errors. Either by itself would have presented problems. Together, they were catastrophic.

At the breach on the west side of the London Avenue Canal (the "clubhouse" breach, as we called it), the sheet piling is minus-10

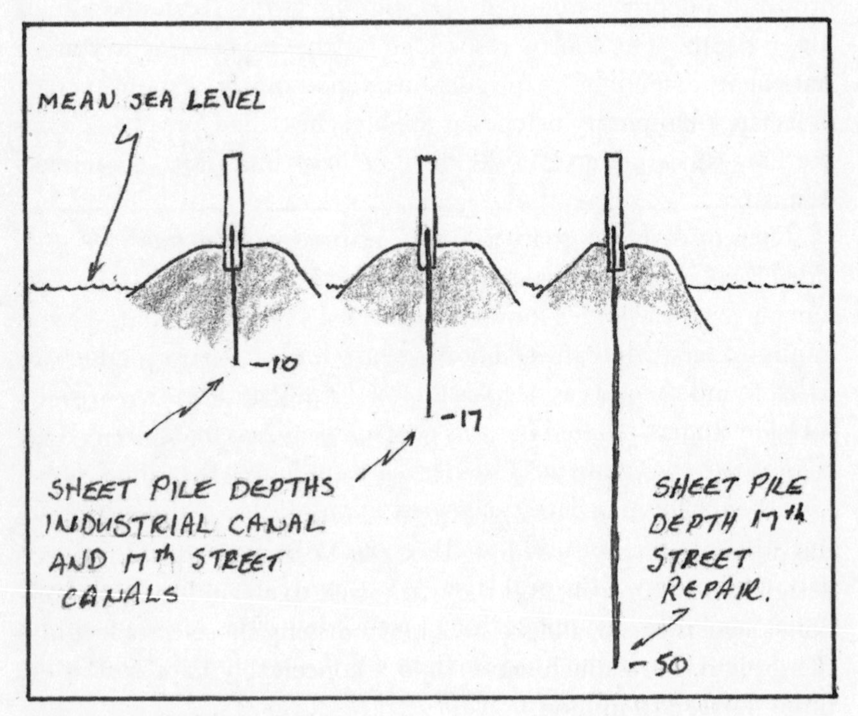

Relative depths of sheet piling.

feet—right to the top of a peat layer—this according to the design memorandum. An enterprising reporter found a consulting engineer for Pittman Construction, the company that lost the lawsuit,

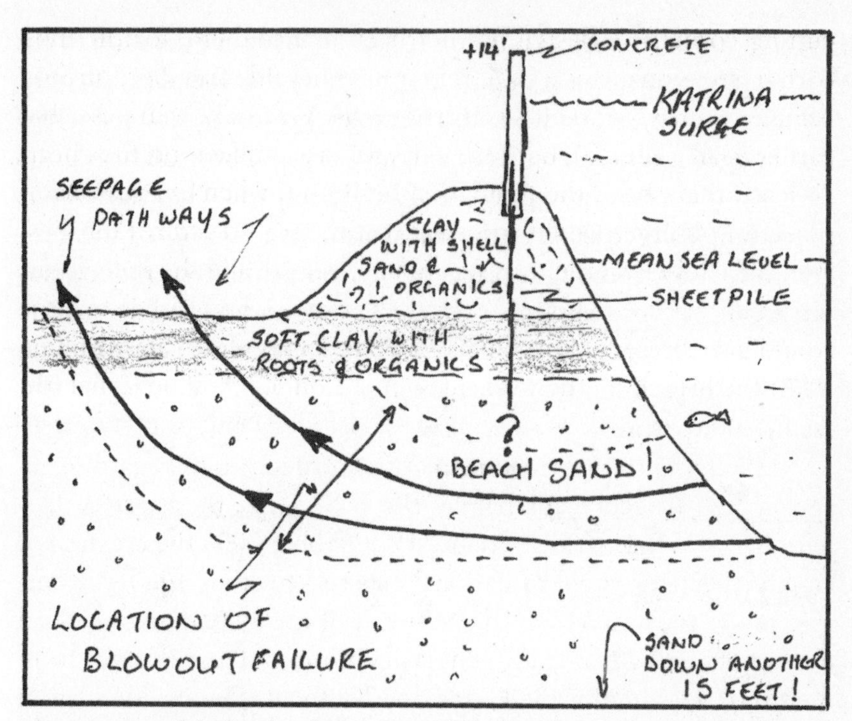

A cross section of the London Avenue Canal/Mirabeau breach showing the soils and predominant failure mode.

who suggested that pilings driven just ten feet deeper might have made all the difference. I don't have the impression that that was a deeply "engineered" statement, but it's quite possibly correct anyway. What a crushing thought. There's no doubt that piling driven fifty feet deeper—to a depth of minus-60, say—would have made all the difference. Still a crushing thought, because the need for such design was easily seen, and the work itself easily and inexpensively done. Steel is cheap, and so is the job of driving those extra feet into the ground. How much extra could it conceivably have cost at the time? *Maybe* $10 million.

There's more. The original design memorandum for the wall at the Mirabeau breach on the London Avenue Canal called for the superior T wall design with sheet pilings driven to minus-26 and the letter A-type batter piles to each side. What we got instead was an I wall design, and our one sonar reading in this area suggests that the sheet piling bottoms out at minus-10. Who made the decision to cut corners from this robust design to something that failed catastrophically? And why? Ray Seed, the Berkeley professor, has suggested that budget cuts may have strapped the Corps. "They still have good people—what they have is less of them," Seed said. "It strikes me as potentially dangerous over the long term." He also noted the general shift away from big projects to environmental restoration, plus the trend toward handing over soil boring and testing to private contractors. Donald Basham, chief of engineering and construction for the Corps, replied, "We have not completely contracted out technical expertise."

Responding to these and many other provocations regarding an agency that is technically part of the U.S. Army, Secretary of Defense Donald Rumsfeld announced in October 2005 the creation of an "independent panel of national experts" to study the levee failures under the auspices of the National Academy of Sciences. So, a *fifth* investigating team. This one will use data and conclusions from the teams working for the Corps, the National Science Foundation (the Berkeley group), and the ASCE. Its report will be peer reviewed. This is fine, but I am not sure what anyone else can say. The evidence is very clear. When we have levee walls sliding forty feet laterally there is no appropriate description other than "catastrophic structural failure." Whether the first water there and at the other breaches seeped through the peat, or down and around the sheet pile, or through deeper sands is almost academic. Important for the investigators, but academic. The bottom line is that these levees were underdesigned and failed under winds and surge equivalent to minimal Cat 1 conditions, much less the Cat 3 conditions the Corps claims they were designed to withstand.

On Wednesday, October 26, Colonel Richard Wagenaar, the district engineer for the Corps in New Orleans—the local boss—paid me a visit in my office at LSU, a meeting he'd requested the previous week. A tall man with short hair, he had two other officials with him, and he got straight to the point. How can we sort out our relationship with the LSU people? I explained the history, and described the opposition we had encountered from the day we first stated that the overtopping scenario was transparently wrong. I told him how we'd had to bring in the big guns from the media (Michael Grunwald with the *Washington Post*, specifically) to listen to our thinking and go to the Corps with the story. I told the colonel that we were thinking of the people of Louisiana, trying to understand the causes of the flooding and the impacts. We don't enjoy being on opposite sides of the table, I said, but we felt the need to get the information and the investigation out in the open. So does the state of Louisiana. I explained that if we had wanted to be completely anti-Corps we could have made much more of an issue of the breaches at the Industrial Canal.

The levees along the east side of this large navigation canal, the ones that had failed and flooded and basically destroyed the Lower Ninth Ward, had been overtopped at about 6:50 A.M. Monday morning, but only for three hours, and by only a foot and a half of water. This water began to erode the outside of the earthen embankments (the scouring effect that is so absent on the inner-city canals). The head in the canal pushed four sections of the flood walls outward. As they tilted, cracks developed at the base on the canal side, with water percolating down and then under the sheet piling. At these breaches the levee was built on a layer of organic marsh and peat with very soft clays ten feet thick—the dredge material from the original 1920 excavation of the canal. Like the soils at the drainage canals, this is a very weak medium. Fifteen feet below the top of the embankment is a layer of soft clays with silt and sand lenses—more poor material. Based on the design memorandum that we found in the files of the DOTD, the pilings at the breach extended to a depth of ten feet below sea level. (This we knew from the exposed sheet piling in the breached areas, but we checked elsewhere with our sonar testing.) The canal is dredged to a depth of thirty-six feet below sea level. Thus there was a linear depth of twenty-six feet of canal that was not blocked by sheet piling, allowing for a potential lateral flow of water under the pilings from the canal. Local residents adjacent to the canal had often said that their backyards were at times quite wet, even during dry spells. This suggests serious seepage from the canal in the best of times. At the very least, this sheet piling should have been sunk to minus-70 feet. Every engineer in the region knows that this wide, deep canal is on the receiving end of all the water and pressure coming out of the Funnel, which I derided earlier as one of the world's best storm-surge delivery systems. This woefully insufficient steel could not hold the static loading from the surge, nor was it deep enough to form an effective hydrologic barrier against seepage below. Around 7:45 A.M., a little less than an hour after the overtopping began, the levees at two sections along the eastern side of the Industrial Canal succumbed to the enormous pressure, the scouring, the substandard soils, and the insufficiently deep sheet piling and failed explosively and catastrophically.

The geotechnical collapse was aided by the overtopping and the scouring, but it was not caused by it. Nor was it caused by the barge that ended up inside the failed section. (In the reflooding of the Lower Ninth from Rita, the barge moved again and landed on the front of a small yellow school bus. Six months later it was still there.) A cursory examination showed that something had knocked nine inches off the top of about forty feet of the southern sections of the concrete flood wall. Initially the Corps speculated that the drifting barge did this while the wall was still upright and intact, but closer inspection revealed that this segment was damaged after it had already tilted to a 45-degree angle. The wall failure was prior to the impact from the barge, which floated into the breach well after the major failure.

In a way we were giving the Corps a bit of a break regarding the Industrial Canal failures and the terrible flooding and loss of life that resulted in these eastern neighborhoods. This design and construction were every bit as bad as on the drainage canals in the Orleans Metro Bowl. I told the colonel this. When his turn came the colonel explained that no one in Washington really understood the extent of the problem here. I got the impression that he was isolated and frustrated with the lack of progress and the lack of cooperation from Washington. He was having to clean up a mess that he hadn't caused with little help from his superiors, who wanted to ignore it. He used his hands a lot as he talked. I felt a lot of emotion coming from him. He was really expressing his feelings. I understood that he had found himself in a very tough situation, newly arrived in Louisiana—a month before the disaster—from an assignment in Korea. There was no way he could have been up to speed on the status of the Corps's work in just New Orleans, much less throughout his region. His background expertise is in environmental science and forestry. I pointed out that we at LSU have bureaucracies just like everyone else, and that he had my commitment to work together.

Of course, I emphasized the problems with the data—how we'd been battling to get the basic facts that the Corps had at its fingertips. The other investigators were getting hundreds of boxes of stuff, but we had to go through the media. I even had a new example handy. Just the night before a reporter with the *New York Times* had directed me to a Web page that the district uses to solicit bids from contractors. One of these bid documents had the boring data that we'd been trying to get from the Corps for two weeks! Without any communication from the Corps, how were we supposed to know about it? I assumed that someone in the Corps had tipped the *Times* reporter. Wagenaar knew about some of these problems, because he had in his hand an e-mail in which Paul Kemp had let fly his frustration about the FOIA request and everything else. He had even brought with him the water-level data Paul had requested! He invited me to call him directly with any other problems.

I wanted to make sure the colonel knew about all the goodies that we could contribute to the cause—the surge modeling, geotechnical work, water sampling, sediment sampling, and mortality studies. What we weren't doing ourselves, we could find. I don't think he had known all of that. I couldn't resist the opportunity to present my spiel about the need for a Gulf Coast Reconstruction Authority, a new Tennessee Valley Authority–type entity headed by a czar empowered to make decisions—informed decisions, advised by scientists and engineers who really know Louisiana. I wouldn't go with the National Science Foundation or the National Academy of Sciences, because they don't know Louisiana. How could they? But the Army Corps of Engineers does know Louisiana. It's really the only game in town. If the Corps would join us we could reach out to the governor's office and the congressional delegation and bring some of the environmental groups on board and make something happen. What about the mayor? Wagenaar asked. Yes, someone from the mayor's office must be close to the action. He said he would contact Major General Strock in Washington about this idea.

So far, so good. I thought it was a productive meeting, and we proceeded down the hallway to meet with the dean of the College of Engineering and the chair of Department of Civil Engineering. I immediately got the feeling that the dean thought that Wagenaar had come to LSU in order to raise hell about us. I tried to correct that false impression. I'm not sure I succeeded. Wagenaar said that he and I had tried to sort out the disagreements—not quite the way I viewed the meeting, but so be it—and that the meeting had been fruitful. Anyway, we'd see what happened next—which happened to be a phone call from Brigadier General Don Riley at the Pentagon, who had been briefed by his colonel about the meeting at LSU. The general apologized for my banning from the meeting in early October. No problem, I said. Let's move forward. Some of the Senate staffers whom I'd met in New Orleans would later suggest that this call from the D.C. brass was most likely because they knew I was going to testify before the Senate Homeland Security Committee on November 2. I wanted to believe instead that the colonel and the brigadier general were really interested in moving forward in Louisiana, not just engaging in damage control.

In the days before the Senate committee hearing, I heard reports that the team representing the American Society of Civil Engineers in the field was being urged by the ASCE to go slowly. Then I heard talk from some ASCE folks about "malfeasance" on the part of the levee contractors, later repeated in testimony to the Senate committee by the Berkeley engineering team. At the hearing itself the malfeasance discussion did seem, to my mind, to concentrate on the contractors. When it was my turn I said that the one breach at the 17th Street Canal and the two at the London Avenue Canal had been geotechnical engineering failures, period: bad soils. The same conclusion can be made for the Industrial Canal levees, I said, although surge overtopping no doubt enhanced their collapse. Regardless, most of the flooding of New Orleans was due to "man's follies." Society owes an apology to those who lost their lives. It owes an apology to those three hundred thousand who lost their livelihoods and their possessions, and it needs to step up and rebuild their homes and compensate for their lost means of employment. New Orleans is one of our nation's crown jewel cities. Not to have given the residents the security of proper levees is inexcusable.

The next day the malfeasance suggestion got all of the headlines. This surprised and disturbed me, and I sent around a note asking reporters to make sure that the malfeasance argument did not distract from the larger point. Even assuming, as the Berkeley engineers testified, that contractors or even individuals may have shortchanged their responsibilities, were such actions the main reasons the levees failed, or did they fail because of much more basic errors of conception and design? According to the Corps's design manuals for the levees on the canals, a safety factor of 30 percent was built into the calculations. That is, the design should have withstood forces 30 percent higher than the highest anticipated. That's a joke. They were doing well to handle forces 30 percent *lower* than anticipated. Both our Billy Prochaska and Bob Bea of the Cal Berkeley team have determined that the factor of safety in the levees is less than 1. Simply put, it had a very high probability of failure. It was a disaster waiting to happen, which is what I told the Senate committee in early November. But even the 30 percent safety factor is actually low for any dynamically loaded structure exposed to rapidly shifting forces such as storm surge; for a structure built on questionable soils; for a structure defending the very life of a major city and the hundreds of thousands of individuals within that city. That low safety factor caught the attention of every investigating engineer.

And all to save a few bucks? If not this explanation, what? We're not going to stop investigating until we know the answer, so it doesn't happen again.

The following week, Friday, November 11, two LSU vice chancellors called me into one of their offices and asked that I henceforth pass all media requests through the university's public affairs office. They told me that my critical remarks to the media about the levees were hurting LSU's quest for federal funding across the board. They also said the university had received complaints about my presenting myself as an engineer and talking about engineering issues. Since I am on the faculty of the Department of Civil and Environmental Engineering, I can certainly see how a reporter could assume that I am one, but I have never so claimed, and I always correct anyone who seems to have that misimpression. Moreover, I said, my nonengineer status did not seem to worry the College of Engineering at Rice University, for example, where I'd been invited to talk. No issue with those folks about my talking as a geologist about geotechnical (soil) issues and foundation failures. Other engineering departments had also requested such talks, and a number wanted to team up with our LSU group to apply for funding, with me directing the effort. In short, I said, I hadn't been made aware of any academic institution having a problem with the scope of my activities.

The gag order, as I referred to it, created a few problems with reporters, who now had to obtain permission to talk to me. They thought this was pretty silly, and so did I. Naturally I objected to the university and asked for clarification. The following Wednesday it backed off and set me free.

At this writing the five investigations of the levee failures in New Orleans are still under way. Various preliminary findings have been released, but the definitive conclusions will not be available until well into 2006. That said, we know the broad outline of what happened in each case. No report will contradict them. At the three breaches on the 17th Street and London Avenue canals, overtopping was not a factor at all, as the Corps eventually admitted, so these failures did not reflect the scouring that aided (not caused, aided) the failure of the Industrial Canal levees. Otherwise, the mechanism of the catastrophic structural failure was similar in each case.

In December 2005, we learned that the semiannual inspections of the levees by the Corps and the various levee boards and state agencies had consisted of a five-hour tour of the levee system. In the *Livingston Parish News*, my hometown newspaper, the headline on the story about this inspection read A DRIVE-BY SHOOTING OF NEW ORLEANS. In the *Times-Picayune* one subhead read: "Lunch, not levees." Elsewhere I saw a sneering reference to beignets and coffee. These attacks were perhaps unfair as they related to the levee boards and the state agencies. From the ground the levees looked fine, given what they were. The problems were with basic design and construction, and the problems were underground. No visual inspection was going to reveal them, and they were *not* the responsibility of the local boards or state agencies, who had no good way to even know about them. They were the responsibility of the only agency that had designed them and then supervised their construction: the Army Corps of Engineers.

This fact was driven home hard right at the end of the year, when the *Times-Picayune*, under Bob Marshall's byline, published incendiary documents showing that a design review in the Corps's Vicksburg, Mississippi, office had found errors in the New Orleans District's assessment of soil strength for the 17th Street levees, errors that had led in turn to "overly optimistic" conclusions about the overall strength of the design. As the *Times-Picayune* reported, this had been discovered in 1990, a couple of years before the concrete flood walls were added to these levees, plenty of time in which to make changes. In fact, the Vicksburg officials directed the local engineers in New Orleans to make necessary changes. The *Times-Picayune* pointed out that Eugene Tickner, then chief engineer of the New Orleans District, cited "engineering judgment" and overruled those instructions. Vicksburg backed off. (It remains to be seen exactly how these documents correlate with the others about the erroneous cross section.)

In a letter to the New Orleans District dated August 8, 1990, the chief of the Engineering Division in Vicksburg questioned the designed depth of the steel-sheet piling. Noting the critical nature of the project and the proximity of the adjacent canal, he questioned the validity of the calculations in the lateral-stability analysis—that is, will the wall fail due to the lateral pressure of the water when the surge comes into the canal? This question didn't surprise me. Billy Prochaska had already shown us that the soil strengths used in the design of the 17th Street levees were too high. The designers had averaged the data for weaker and stronger layers, a no-no in the geotechnical engineering of levees. Where the pertinent 1982 borehole data had strengths as low as 100 pounds per square foot, the designers had used strengths as high as 320 pounds per square foot. Standard practice in levee design is to model based on the soil strengths in the weakest layers. Weakest layer, weakest link, common sense. Also, the New Orleans District used soil data from locations over a mile apart. In this part of Louisiana, where the land is the product of delta switching and channel growth and abandonment cycles, one after the other, one on top of the other, we can have a dramatic change in the geology in one mile.

To my mind, these Vicksburg documents are the worst of the smoking guns in the levee investigation to date, although I won't be surprised by anything that follows in the months and years to come. It's sickening to think about this, especially since our analysis at the Hurricane Public Health Center shows that almost nine hundred of the deaths and the great preponderance of the property damage in New Orleans and Louisiana as a whole were in the neighborhoods flooded by these drainage canal flood wall failures.

Toward the middle of February 2006, the *New York Times* informed us that the Corps was spending more than three hundred thousand dollars to create a 1/15-size physical model of Lake Pontchartrain and environs (basically a pond such as kids sail model boats on in parks), where they intended to simulate the waves and surge during Katrina. Well, good luck. I have worked with such scale models in South Africa and there are real problems with "scaling" water depths, waves, and so on. How do you hope to create a revolving wind field such as a hurricane? I really don't know what the Corps hopes to achieve here, even though it is spending almost twice the money our whole team had for all its levee work. As the *Times* article points out, the process was immediately attacked by critics as an expensive attempt to deflect the blame from the Corps.

ELEVEN

NOW OR NEVER

I guess the acronym was pretty good: the PELICAN Commission, for Protecting Essential Louisiana Infrastructure, Citizens, and Nature. Otherwise, the legislative initiative I came to think of as the Pelican Brief (for two reasons: John Grisham's bestseller about environmental damage and because its effective life of about twelve hours) was a terrible mistake for the state. Let's hope it doesn't set the standard for the thinking that will guide the rebuilding and restoration in the years to come. If it does, we're probably sunk.

My part in this fiasco began on September 14, with an e-mail from John Barry, author and fellow traveler on the major questions concerning what had happened and now needed to happen in New Orleans. John was part of a team being gathered by Senators Mary Landrieu and David Vitter to develop a plan for the restoration and reconstruction of Louisiana that would be the basis for the main federal support legislation. Specifically, John had been invited to be part of the working group focusing on flood protection and restoration of the wetlands; the deadline for the recommendation was one week away. I immediately phoned and asked John if I could be part of this exercise. When I had left the Department of Natural Resources almost exactly ten years earlier, I wasn't invited back to the table, but I thought that someday I might be. This could be a step in that direction. Who wouldn't want to be involved here? These decisions could be critical. The science could be critical. I wanted to bring expertise to the table. I wanted to push hard for my belief that the catalyst to compromise is a thorough understanding of the science. John then asked if I wanted to chair the working group! No way could I take on that responsibility, not with the levee investigation heating up, but count me in otherwise. John ended up chairing it himself.

I missed the first conference call—that same day, the fourteenth—because I was in the field at the levee breaches helping out a *National Geographic* team in the morning and then a French documentary team in the afternoon, which turned out to be one of the saddest moments in the whole catastrophe for me. We went to the Upper Ninth Ward, which was just about dry (this was before Rita flooded it again), with the body search just getting under way. We searched for a site at which to film the last interview, which was to be about my personal feelings, and we did not have to go far to find a totally eerie one: trees and power lines down, some buildings wind damaged, a thick, black-brown, stinky mud underfoot (one of the vehicles got stuck along the way), and the buildings all absolutely black from the mud and with a ring of oil near the top of the floodwater line. We stopped between a small Baptist church and a midsized family home. This home showed no sign of damage—this was the eerie part—and the body-search groups had not come here yet, so the front door and windows were intact. There were even hanging flowerpots on the front veranda, though the plants, like those everywhere, were black and dead from the saltwater and the oil. The flood line was just below the eaves. It struck me immediately—this home has a rich history. Where were the occupants? It was obvious that they had evacuated. Was one of these dogs wandering around their children's pet? My heart was so very, very sad. I felt defeated. Could we have better predicted what had happened in Katrina? No, we couldn't have and academia had not designed or built the levees. That night we said some extra prayers for all those families.

A National Guardsman in a Humvee rolling past asked, "Did you see a body back there? We have a report of a body." No, we hadn't, but the sweetish smell of death was in the air, and I was sure that they would find bodies the next morning as the search crews got started. In the months following, I passed that home many times and never saw any signs of life. (Six months later, still no signs of life.)

That evening I told John Barry about my day in the Upper Ninth. In addition to the science, such images and hands-on, day-to-day experience in New Orleans were what I could bring to the committee. No one else in the group had that perspective. John said he had tried during the conference call to alert the assembled team about these issues, especially as they involved the Corps, but he didn't feel as though he had had any impact. Then he rang off to watch President Bush's speech from the stage in the French Quarter. (I read it instead.) Later that night I received the "welcome aboard" e-mail that had gone out from one of Senator Vitter's staffers to the participants in the call, and I concluded that I was an official member of the team.

The following morning I was leading a group that was flying over all of the major levee breaches, both in the city and out to the east in St. Bernard and Plaquemines parishes. I brought along the cameraman for the French crew, because we had a deal: In exchange for the flight he'd provide me with a copy of that high-resolution footage, which is much better than any of us could achieve with our handycams. But I therefore missed the second conference call of the Pelican working group. That night I studied a report submitted to the group by the Association of Flood Plain Managers. "No adverse impact" is one of the goals of this organization. This principle emphasizes the importance of understanding how natural systems function and then defining, at the local level, how best to manage rivers and drainage systems. In essence, "No adverse impact" floodplain management tries to assure that the actions of one property owner do not adversely affect the rights of other property owners, as measured by increased flood peaks, flood stage, flood velocity, and erosion and sedimentation. This is definitely the right idea. Given that flood damages in the United States continue to escalate, now approaching $6 billion annually, even with the billions of dollars spent for structural flood control and for other structural and nonstructural measures, we really need to change the way we do business. This association has some great ideas and ideals, and their report was a serious piece of work.

I was pretty encouraged, but when I talked with John later that night, he repeated his unease that for a second time he had failed to get his group to hear his concerns about the existing levee system, the multiple failures, the probable causes, and the implications they held for all future plans and projects. It seemed pretty basic to him—and to me, of course—but he wasn't getting through to them. He wanted me to try on the next conference call, but first he wanted me to contact Garret Graves on Senator Vitter's staff, who would actually draft the plan. It was late, but I phoned Garret and presented a brief version of our early concerns about the levee breaches and what I thought they should mean to this team. He heard me out, but he was also tired. Understandable. It was 1:00 A.M. in Washington. He had to interrupt the call, and said he'd get back to me, but did not. After half an hour I decided to put all my ideas and concerns down in an e-mail to the group.

I've said before that I can be naive. I should have paid more attention to a note John had sent me after our conversation but before I had called Graves. This is a group of Hill staffers and lobbyists, John said, and their approach seems to be "piecemeal," a compilation of old Corps proposals, and not a comprehensive reassessment. If I'd thought about what John had said and realized we were working with a group of lobbyists, for the most part, my note would have been a little less frank. But I didn't pick up on this salient fact, and my note was not less frank.

I began with a pretty vivid description of the scene in the Lower Ninth. I laid out the issues with the levee breaches and wrote, "I believe it will be a long, long time before anyone trusts the COE [Corps of Engineers]. . . . No one trusts the COE." I mentioned that we (LSU people) had been working with the Greater New Orleans Planning Commission to develop a design for a floodgate for the infamous Funnel east of the city—the confluence of the Mississippi River–Gulf Outlet and the Intracoastal Waterway. No one wanted to wait fifteen years for the Corps to build the project. And as for the state, its agencies do not have the infrastructure to manage the comprehensive restoration and levee work—a conclusion I based on my brief tenure as the main coastal restoration guy. Fear of corruption would also dog any vast program run by the state. Fact of life. I therefore mentioned my idea for a new body along the lines of the Tennessee Valley Authority that could "cut through red tape because . . . there will be no interdepartmental and intradepartmental turf battles." This authority would have a czar and a blue-ribbon panel of scientific and engineering experts and business administration advisers. It would have cabinet-level representation, and it would have the right to seek its own appropriations and independent financing. It would spread the rebuilding dollars to contractors throughout the tristate area (Louisiana, Mississippi, and Alabama). It would pay above minimum wages. It would not allow contractors to use mainly illegal immigrant labor. It would be a twenty-first-century solution to a twenty-first-century problem.

The next morning Garret Graves's e-mailed nine-point plan for a "Marshall Plan–type Commission" was waiting for me and all of the other working group members. This was based on the team's discussion of the previous day. The call for "programmatic authority" for the Pelican Commission was in line with my TVA thinking. The plan called for an independent review of the levee failures and for Cat 5 levee protection. It advocated an "inter-related" approach to "hurricane protection, flood control, navigation, and coastal restoration." It stated that coastal restoration should not be viewed or construed as "eco-system restoration" but as "protection of our coast and energy infrastructure" that is in the national interest. All this was a promising start, but there were also red flags, especially the "navigation" initiatives, like dredging canals and improving ports miles and miles away from New Orleans. How did these projects help restore and rebuild southeast Louisiana? And why wasn't the closure of MR-GO and stopping all plans for building new locks for the Industrial Canal in the main plan? Those are basic requirements for any responsible plan. Everyone knows that. But the point that really got my attention was this one: "Work should proceed in an expedited manner—including the possibility of waiving, reducing or streamlining environmental, cost-share, economic and other considerations." That was radically open-ended. Regarding the environmental "considerations," this statement could lead to weakening wetlands permitting regulations, leaving the door wide open for developers to build more homes or factories in drained wetlands—the very wetlands we need for storm buffering in many cases.

Immediately—prior to the planned 9:00 A.M. conference call—I sent the planning team a list of these concerns. On that call with about a dozen people (as best as I could figure) someone called "Bubba" on Senator Landrieu's staff complimented me on my *Meet the Press* interview, and I thought, great, this is starting off well. Then, at John Barry's prompting, I tried to inform the assembled listeners of my observations concerning the levee failures and the Corps's failure to warn people on the day Katrina struck. I asked about the relationship between dredging harbor canals nowhere near New Orleans and flood protection, wetlands restoration, and Katrina relief. That's when an unidentified stern male voice tried to shout me down. "The oil companies have said they will go overseas to build rigs if we don't dredge the canals!" he yelled. *What a lot of bunkum*, I thought—and then it hit me. These people—some of them—are all about their favorite deals! I started to sense that the two callers who seemed to be supporters of the Corps were getting rather annoyed with me. I could hear the contempt in their voices, so I tried to deflect the antagonism by suggesting that deepening the canals was not my concern, but they had better try to encase it in an economic development envelope.

The discussion then turned to the value of a scientific panel. On one of the earlier calls, the group had discussed using the National Academy of Sciences. An unidentified woman suggested that we should set up a committee of scientists from Louisiana, because they know the area and understand the problems. Of course, I agreed. One of the main reasons we set up Team Louisiana for the levee investigation was to get local engineers with local knowledge of the soils and other issues. (This question of turf and reputation can be surprisingly significant even in science. In many third world countries local people frequently have the attitude that any idea or scientist from the first world, especially from the United States, must be better than their own. It's an inferiority complex of some sort. When I returned to South Africa after my graduate studies at LSU I had to fight the feeling that *I* must understand the local problems better than the local scientists, who had lots of experience. Such nonsense—and we have it in Louisiana as well. Some group from the *National* Academy of Sciences must be better than our people, when quite the contrary is the case.)

Slowly but surely, two guys (unidentified, at least for me) seemed to begin dominating the conversation, and before I really knew what was happening the decision had been made that the Pelican Commission wouldn't need any science at all. Maybe I was the problem, I don't know. I was the only representative of science in the group, I was becoming a squeaky wheel, therefore just get rid of science. That was kind of the way it seemed to be going, and I was deflated. Other than John Barry, the woman who wanted some local scientists involved, and me, no one in the group wanted to take this opportunity to do the rebuilding and restoration right. Instead, they wanted their projects. When someone else rang off the call, I followed suit.

Less than twenty-four hours later, Marc Levitan received a call from Randy Hanchey, deputy secretary of the Department of Natural Resources, complaining about my bad-mouthing of the Corps in my e-mails to the group, which he had. (Hanchey worked for the Corps for many years before his retirement in 1998.) The e-mail's path began with Hunter Johnston, a Washington lobbyist and former senator Bennett Johnston's son—and a member of the Pelican working group. Johnston had forwarded one of my e-mails to Ken Brown, CEO of the engineering and architecture consulting firm Brown, Cunningham & Gannuch. Most if not all of the firm's principals are former Corps employees, and they do a lot of work for the New Orleans Port Commission (beneficiary of the controversial lock on the Industrial Canal), numerous districts of the Corps, and many other others. (A few months later—January 2006—the company received a sole-source contract from the New Orleans District of the Corps to look at drainage in New Orleans.) Ken Brown in turn had sent my e-mail to Hanchey, who in turn complained to my colleague. I was being honest but that seemed to get me into trouble at every turn. I then found out who the other members of our little club were. The *Washington Post* reported that they were lobbyists from such powerful firms as Patton Boggs, Adams & Reese, the

Alpine Group, Dutko Worldwide, Van Scoyoc Associates, and a firm owned by former senator Johnston. According to the *Post*, the group also included a lobbyist for the Port of New Orleans, a lobbyist for Verizon Wireless, and three lobbyists who were former aides to House Transportation and Infrastructure Committee chairman Don Young of Alaska.

After this incident involving a senior state employee with very obvious pro-Corps leanings, and having learned the exact makeup of the planning team, I decided to go directly to Garret Graves. On September 16, I sent him an e-mail expressing my concern that what I considered a confidential e-mail within our group had been disclosed to a private engineering company. I questioned if this action should disqualify the engineering company from participating in the reconstruction efforts, and I deplored what seemed to me like politics and deal-making already at work even in Louisiana's greatest hour of need.

I received no reply. A few days later John Barry sent the committee a note reminding us all that "projects need to be integrated and priorities set. Particularly, ties need to be developed between the construction projects and the coastal restoration effort. It seems that we all are hell bent on getting projects funded (money) rather than a comprehensive reconstruction/restoration plan. It may be important to take at least a ¼ step back for a second, and make sure things are approached comprehensively." John's team would have been well advised to heed their leader's wise counsel. As a longtime Louisiana resident as knowledgeable as anyone about these issues, going all the way back to the 1927 flood, John knows what he's talking about. But his committee went on its merry way—around him. Perhaps John's presence (and then mine) were just meant to be some sort of cover? In any event, less than a week later Senators Landrieu and Vitter introduced their Louisiana Katrina Reconstruction Act, seeking $250 billion in federal monies—that's $250 billion over and above the $63 billion already authorized by Congress for emergency relief purposes.

What were our senators thinking? They had taken all the suffering from Hurricane Katrina and turned it into a boondoggle, a porkfest. The request came to $55,000 for every man, woman, and child in the state (pre-Katrina, even more per capita post-Katrina). Everybody was taken care of in this legislation—every commercial interest, that is: energy (oil and gas), construction, ports, shipping, you name it. The $35 million for the seafood marketing campaign caught the eye, as did millions for a sugarcane research laboratory. A long article in the *Los Angeles Times* on October 10 reported that just five days before the bill was introduced one of the energy companies had retained the services of Lynnel B. Ruckert, Senator Vitter's former campaign manager and wife of his chief of staff. Smart move. The bill made $2.5 billion available to the private utilities.

All of this was catnip for the press, who predictably and quite rightly ate it up. They had great sport with this legislation. The *Washington Post*'s outraged editorial was titled LOUISIANA'S LOOTERS. (The alliteration proved irresistible; others in the media picked up on it immediately.) The editorial said, "This is the equivalent of New York responding to the attacks on the World Trade Center by insisting upon a federally financed stadium in Brooklyn." Then and there the legislation was effectively dead, I'm sure. Reporters and columnists pulled out favorite tales of colorful corruption, Louisiana-style, going back to the era of Huey Long and his brother Earl, who was railroaded into a mental institution by his wife. (I learned that one of Huey's mottoes had been "Share the Wealth," which can be a noble sentiment or something else. Likewise "Every Man a King.") Did a single story fail to mention that former governor Edwin Edwards is right now sitting in prison (as is his son)? I wonder if former senator John Breaux enjoyed being reminded of his quip, delivered after supporting some Reagan-era, trickle-down budget proposal, to the effect that "my vote can't be bought—but it can be rented." Funny at the time, I'm sure.

Mississippi governor Haley Barbour was upset that the outrageous request from his neighbor state indirectly hurt his own request for $34 billion, but in the end, Mississippi Republicans in Washington—let's tell it like it is—ended up getting three or four times as much aid per capita for their state as Louisiana received. (Barbour is a former chairman of the Republican National Committee.) No matter. Not three weeks after Katrina struck, with parts of New Orleans still underwater, with the catastrophe still unfolding, with the whole world watching, it was suddenly open season on Louisiana and its politicians.

The section of the legislation I was most familiar with would have directed $40 billion to the Army Corps of Engineers for their ultimate wish list of projects, some of them, like the $740 million for the new lock on the Inner Harbor canal, long discredited as among the worst of the worst boondoggles. The final bill exempted all of the rebuilding work from the regulatory provisions of the National Environmental Policy Act and the Clean Water Act. This was unbelievable. John Barry's and my advocacy of an oversight role for a scientific panel was not included in the legislation.

John was upset with the final package, and said so. I was really upset, and said so. Up to that point in the Katrina story I had thought that the offer of a paid vacation in Las Vegas to New Orleans cops and others was the worst political/PR blunder. Or maybe it had been President Bush's lame flyover. Or the president's lamentation about Senator Trent Lott's demolished shorefront estate in Mississippi. Or Brownie's admission that he had been out of the loop regarding the thousands of people stranded at the Convention Center. Now I was certain that the Pelican Brief would be the worst by far, because it stood the chance of undermining the entire recovery effort.

On September 1, President Bush had said, "I'm confident that with time, you get your life back in order, new communities will flourish, the great city of New Orleans will be back on its feet, and America will be a stronger place for it. The country stands with you. We'll all do all in our power to help you." As the months passed this was becoming more and more doubtful. On December 15, the president announced that he would ask Congress for $1.5 billion to bolster New Orleans's levee system. His own "best man" for Gulf Coast reconstruction, Donald Powell, admitted that there were design and construction flaws that needed to be remedied, and he added, "The federal government is committed to building the best levee system known in the world." Well, this would cost a lot more than $1.5

billion—about twenty times that much. Time will tell if those funds are really forthcoming, and if they're not we can lay a good part of the blame on the Pelican mistake. In December, Senator Larry E. Craig (R-ID) told a home-state newspaper, "Louisiana and New Orleans are the most corrupt governments in our country, and they have always been. Fraud is in the culture of the Iraqis. I believe that is true in Louisiana as well." Pretty harsh words, but in line with lots of similar comments from Washington lawmakers. John J. "Jimmy" Duncan Jr. (R-TN), chairman of a key subcommittee involved in reconstruction efforts, told the *Los Angeles Times*, "I don't think I have ever seen an issue flip so quickly as this did." The *Shreveport Times* noted that D.C. and state officials agreed that "the tide of support for helping the state turned when [the] submitted legislation seeking $250 billion . . . had little or nothing to do with hurricane protection." Meanwhile, Mississippi's legislators got high marks for working quietly behind the scenes to steer resources to their constituents.

What a blunder. What a loss. There were other blatant acts of profiteering, but this one got the headlines. It was embarrassing, and when the dung instantaneously hit the fan, the two senators' apologists hurried to explain that the final sum would probably be in the range of $200 billion, not $250 billion. This didn't help. Senator Landrieu's spokesman said to the *Los Angeles Times*, "Standing up the region's economy will help stand up the American economy. The lobbyists and the entities they represent tend to be among the most experienced experts available who have direct real-world knowledge of the situation. They are advocating for a position and for a client, but usually from a vantage point of expertise that can be very beneficial to us."

Who, exactly, is "us"?

Throughout the fall I and others with LSU and elsewhere were working on two fronts: continuing to investigate the levee failures that had caused this catastrophe, and developing political support for the ideas that could prevent another one (while hoping that the Pelican Brief hadn't made this impossible). The truth about the levee failures and the plan for the future are absolutely related. Most obviously, as I wrote in my note to the Pelican planning group, I don't believe we can entrust overall control of the incredibly important restoration work to the Corps. Not now. I had my chance to lobby Governor Blanco on this point and many others in December, a couple of days before she briefed a House committee on the situation in her state. This was my first conversation with the governor one-on-one, and she was great: businesslike but warm and friendly, and I have no doubts about *her* motivation or her grasp of the issues. We met for two hours, even discussing my problems with LSU, some of which I've sketched in these pages.

I pressed the critical point that there is only one way to describe the levee breaches, and that is "catastrophic structural failure." I was pleased she used that exact phrase before the committee. Clearly some congressional Republicans are trying to blame the state and local governments for the catastrophe; just as clearly the responsibility lies with the Army Corps of Engineers and the federal government, which had, through its agency, failed to deliver protection from even the modest storm that Katrina was in New Orleans. But even if Katrina had been the Cat 4 storm the Corps portrayed, its excuse that Congress hadn't provided enough money to protect against such a storm just does not hold water, if you'll accept the pun. Switching metaphors, the squeaky wheel gets the oil, and all the Corps had to do was complain, complain, complain, help the state complain, and get citizens' groups to complain. The Corps doesn't seem to have much problem getting the special-interest lobbyists to procure the funding for canals and harbors. If it knew it had such an inferior product in the levees, if it could not supply the level of protection required by the 1965 Flood Control Act with the funds provided, then it should have organized a public information campaign to tell the public that they were living with a ticking time bomb.

I suggested to Governor Blanco that the Corps would have to be involved in the new flood-protection plan, whatever it turned out to be, because there is no way to push this behemoth aside entirely, but it could not have final authority. The state needs to take the initiative and design the best system for the future. In the three months since I'd started pushing my hope for a TVA-like federal reconstruction authority, I'd realized that that just isn't going to happen. That idea was DOA, I'm afraid. What *is* possible is independent state leadership, with federal funding—convincingly independent, to address the state's image problem, now even worse, thanks to the Pelican Brief. The governor had already set up the Louisiana Recovery Authority (LRA) to be the legal conduit for whatever federal reconstruction money Washington still feels like dropping in the can after the Pelican Brief fiasco. The LRA is modeled on the Lower Manhattan Development Corporation that monitored the work at Ground Zero and did, by all accounts I've read, a great job, in no small part thanks to the system of "integrity monitors" put in place by Mayor Rudy Giuliani, private-sector watchdog firms assigned to every construction company involved in the cleanup. As a result, there was not one significant corruption scandal in that effort. This is incredible, given New York's own reputation for funny money in the construction trades. Governor Blanco did not copy this oversight scheme, but she did institute stringent controls for verifying the disbursement of every federal dollar, to the point of irritating some communities waiting for FEMA checks. At a special session of the legislature, the governor also created a twenty-one-member Coastal Protection and Restoration Authority, itself somehow augmenting the old Wetlands Conservation and Restoration Authority, charged with devising a master plan for flood protection and coastal restoration. But the $32 billion for flood protection and wetlands restoration, advanced principally by our old friend Randy Hanchey, was really just a dusted-off version of the Corps's old plans. It even bore a strong resemblance to the $40 billion allocated to the Corps in the ill-fated Pelican legislation. I say throw out the old and start with a fresh sheet of paper. Otherwise we are doomed to repeat the mistakes of the past.

Almost every local and state (but not federal) official has lined up to proclaim that protection from a Cat 5 storm—not just a promise, but a program with guaranteed funding in place—is mandatory for luring people and businesses back to the city. Various Corps officials have said all along that such an upgrade would take a minimum of ten years, perhaps fifteen, perhaps more. Congress has already bestowed $8 million on the Corps to study how to do it, with a preliminary assessment due in the spring of 2006, and a final plan due eighteen months later. Randy Roach, the mayor of Lake Charles, which was badly damaged by Hurricane Rita, said in anger about all such planning, "We're burning daylight. We have just a matter of months, not a matter of years. We have studied this. I don't want to hear about a new study." Indeed, a small group of independent engineers and scientists—no politicians, politically connected members, lobbyists, contractors, or anyone with any financial or political stake in the work—could sit down and in one week craft an ironclad plan for saving southeastern Louisiana. Whoever selects this panel would have a few dozen men and women from which to choose, and any combination would come up with essentially the same plan. This is not rocket science. It never has been. It requires only a political system that can get it done. The plan they come up with will cost $30 billion. It will require ten years to complete. As a political reality, the Corps will build the projects, but at least it can be forced to build them to the state's specifications and overall plans. We need this thoughtful, comprehensive, big-picture plan that can achieve the goal with a minimum of lost time and wasted dollars. Politically, we need a clear, effective mission and purpose that emphasizes protection for the people. Emphasize protection from flooding and the restoration of the wetlands will necessarily follow. But protection must drive the process, politically. This is the way to confront Speaker of the House Dennis "A lot of that place could be bulldozed" Hastert and others who question whether low-lying, high-living New Orleans is even worth the trouble to rebuild.

(If Hastert's remark was a calculated trial balloon, it went over like a ton of lead. He caught a lot of flack, but Randy Lanctot, executive director of the Louisiana Wildlife Federation, of which I am a card-carrying member, sent around a note suggesting that the Speaker's basic question about rebuilding, if delivered in the right spirit, was legitimate. Maybe it just reflected the "practical way Midwesterners think," Lanctot added hopefully. Hastert is certainly aware of the vital role played by the ports in Louisiana for the grain producers in his own state of Illinois. *How* do we rebuild? Should the levees be upgraded to Cat 5? Should the lowest levels of the city be raised? To use the buzzword, shouldn't rebuilding be "sustainable"? Recipients of the e-mail agreed that these are the right questions, but they weren't willing to give the benefit of the doubt to Hastert's motivation. Neither was I.)

A few days after our conversation Governor Blanco was off to testify in Washington, D.C., and I was off for the Netherlands to see for myself what the best flood-protection system in the world looks like. The trip was sponsored by WWL-TV, the CBS affiliate in New Orleans that never went off the air and was so important during Katrina. The editorial idea was for me to join veteran anchorman Dennis Woltering, a New Orleans institution with a soft warm smile, to study what the Dutch have accomplished following their own catastrophic flood of half a century earlier. Despite my Dutch heritage I had never been to my ancestral homeland. As a child I had heard how the Netherlands had "turned back the ocean" and "captured the land from the sea," and along with every other kid, I guess, I knew the story of the little Dutch boy who discovered a hole in the levee and saved the day by plugging that hole with his finger. Today Dutch engineers are considered just about the world's leading river flood-and-surge protection experts. I was very anxious to see for myself what they have.

Our host for the week was Jurjen Battjes, a recently retired engineering professor who had been a member of the levee investigation team assembled in New Orleans by the ASCE. Jurjen is tall, pleasant, and very knowledgeable, and he has a wonderful twinkle in his eye. It was Jurjen who asked when he first surveyed the drainage canals in New Orleans with their pumps inside the city, "Why invite the enemy so deep into your camp?"

The Netherlands is about half the size of Maine. Almost a quarter of its land is below sea level—the lowest land 22 feet below—and without the protection of dunes and dykes about 65 percent of the country would be inundated either by storm surges from the sea or by floodwaters from the great rivers (the Rhine, Maas, and Schelde). The vicious winter storm in 1953 resulted in 1,800 Dutch deaths, destroyed 4,000 buildings, and flooded 625 square miles. (Katrina flooded about half as much land.) Such a catastrophe demonstrated the woeful insufficiency of the protection system in place at the time, and the nation mobilized practically overnight to set things right. And how. They have contrived an engineered landscape of flood control structures—dunes, dikes, dams, barriers, sluices, and pumps—that puts to shame anything Louisiana has to offer. The contrast is embarrassing, frankly. The system is now an integral part of Dutch national pride, and rightly so. It demonstrates what determined and highly motivated people can accomplish. It is a standing challenge to the citizens of Louisiana and the United States.

As Jurjen explained to the visiting Cajuns (including one honorary Dutchman), the local policymakers and engineers looked at their old, failed system and presented the nation with a choice: rebuild the existing levees to higher and stronger standards (they had suffered seventy-eight breaches in the big flood), leaving a certain vulnerability, or start all over, in effect, and think really big and outside the box. The Dutch chose the latter option. Cost was barely an issue, because the life of the nation was at stake. (I write that sentence with a heavy heart. With the life of the city at stake in New Orleans, money was always invoked as the reason for not doing this or that.) The complete overhaul would be entrusted to one central agency. Local authorities could be charged with maintenance duties, but overall design and construction would be a national responsibility. One very interesting aspect of the Dutch system from which we could take a lesson is that the design of certain of the larger structures was set up as a competition, with engineering and contracting companies challenged to submit their own designs. The planners believed this was the best way to utilize all available brain power and creativeness. (The motto of our Army Corps of Engineers is *Essayons*, French for "Let us try." Maybe it should now be, in Louisiana, *Maintenant, laissez un autre essayer*, or "Now let someone else try.")

The first and most important decision regarding the new design was prompted by the same problem Jurjen had pinpointed with the drainage canals in New Orleans. The former levee system, long and sometimes sinuous, had allowed surge water to penetrate deep into populated areas. The system had needed to be "shorter." It had needed large barriers across estuaries and other open waterways, thus cutting back on the need for extensive interior defenses. The replacement system has a basic, outer line of defense backed up by a secondary defense of compartments within compartments. Experts in probabilities and risk management determined just what degree of protection was needed where. Highly industrialized areas now have protection against anything short of the ten-thousand-year flood, while rural farmland is more vulnerable. (In New Orleans, the Corps thought its standard project hurricane was, in essence, a two-hundred-year storm.) As in Louisiana, the essence of the problem for the Dutch was that the draining of their peat bogs and marshes initiated the vicious cycle of subsidence, ever more subsidence. Understanding this, their system now allows nearly normal tidal interchanges in some locations, while others are carefully monitored and, if necessary, pumped in order to maintain groundwater elevations, so the decomposition and subsidence rates are held to an absolute minimum.

The most vulnerable and critical link in the Dutch defenses is a channel system known as the Hollandse IJssel that penetrates deep into highly industrialized and settled areas. Less than five years after the flood, they had designed and built a double storm-surge barrier in this channel consisting of vertical sliding gates three hundred feet wide, a ship lock, and a road bridge. These gates, as well as those elsewhere, are beautifully designed with two of everything, in most cases: two surge gates, two sets of locked doors. Of course, some commercial and recreational craft, especially sailboats, have high clearance needs, so these gates feature a narrow section that can be raised, in the manner of a drawbridge. I can see such a structure on the Industrial Canal just off Lake Pontchartrain.

One of the highlights of the trip was the visit to the Ooster Schelde barrier, one of five other major barriers known as the Schema Afsluitingen. The Ooster Schelde is a series of 62 sluice gates, each 120 feet wide and separated by man-made islands. The whole structure is 4 miles long and allows free tidal interchange with the estuary, keeping intact the natural functions and the commercial harvesting of seafood, but the gates can be closed quickly, thanks to their ram-driven hydraulics. Cost: about $4 billion. This barrier and the four others with similar functions face the North Sea, whose tidal range of nine feet and potential exposure to huge waves necessitates far more robust structures than we would need anywhere in Louisiana.

The most impressive and famous of the structures in the Dutch system is the storm-surge barrier at Maeslandt on the extremely busy Rotterdam Waterway. This design was also the product of an open competition. The main element of this barrier is two enormous "butterfly wings," one on each bank of the channel. Each wing looks something like two oil rigs joined at one end by a gigantic

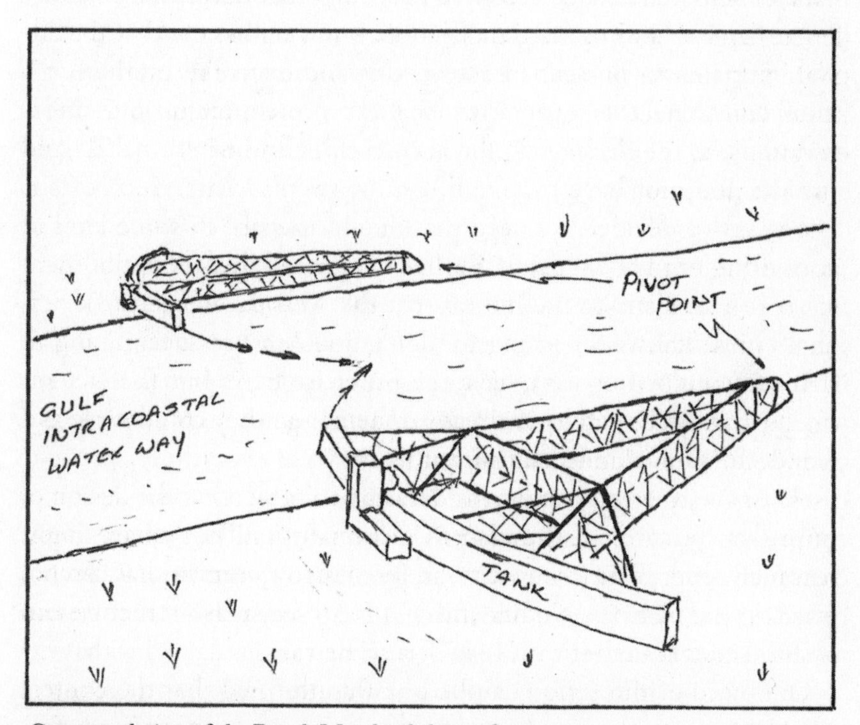

Conceptual view of the Dutch Maeslandt butterfly surge gates as they could be built in the Gulf Intracoastal Waterway at the location of the Funnel.

ball-bearing assembly to form an enormous V as long as the Eiffel Tower is tall. The open end of each V is a large water tank. In normal times each wing lies beside the channel flat on the ground, so to speak, and completely out of the way of all maritime traffic. Called into action, they swing out and meet in the middle of the channel. The two large tanks are then filled with water and sink to the bottom, sealing the waterway from any conceivable storm surge. Otherwise, ships have an unimpeded passageway at all times. The whole system is computer-driven, with backup redundancy for every mechanism. It is an amazing sight, truly an engineering marvel—and it is something we could easily adapt and build in Louisiana. We have some of the top riggers and shipbuilders in the nation and therefore the world. Fabrication of such a structure should be a piece of pie for them. And the cost would be only $700 million, in current dollars.

Are we in Louisiana and the United States even capable of such comprehensive, science-sensitive planning, and then performance? It's a shame I have to ask the question, but we all know that the answer remains to be seen. I returned home from the Netherlands more convinced than ever that such Cat 5 protection in Louisiana is very doable, requiring only the political will and wherewithal. And the job does not have to take fifteen or twenty years. Ten years is enough. I've decided we need the Dutch onboard in some kind of consulting or oversight role. Earlier I criticized the susceptibility of some in Louisiana to the notion that the *National* Academy of Sciences must know something we don't, and I'm not succumbing to that same error now. We have the expertise here, but the fact remains that the Dutch have already done it, and they could serve as a wonderful fail-safe mechanism for us.

We have to give careful consideration to subjecting the design of individual features of the plan to a competition, just as architects routinely engage in competitions for design commissions. Such a process worked in the Netherlands—it is one key reason their system is so excellent—and it can work here. The various agencies that expect a piece of the action can be bluntly informed that the control they have always enjoyed is a luxury we can no longer afford. Encouraged by the independence of this process, private foundations would surely underwrite the competition.

We must immediately move forward on two fronts: a major barrier system and large-scale wetlands creation. No more Band-Aids! If this isn't clear by now to everyone concerned, something is seriously wrong in the state of Louisiana. Any big-picture surge-protection system for this state must include major barrier levees and flood gates in conjunction with creating wetlands and barrier islands. The best procedure is to build the hard structures where they can be protected by an existing "platform" of wetlands, no matter how fragile, and where there's a reasonable supply of good soil-building material for levees. The wetland platforms seaward of the barrier levee and floodgates would be the target sites for future wetlands creation projects, with barrier islands seaward of the expanding wetlands base. Barriers—wetlands—barrier islands: This progression assures the survival of the estuarine bays necessary to keep the commercial and recreation fishing industries alive, as well as to supply the breadth of natural habitats that make up this unique ecosystem.

Here are the main elements of the major barrier system:

- At the main entrance to Lake Pontchartrain, build a flood-control structure with multiple gates in the Rigolets along the I-10 Bridge. (The Dutch often combine a bridge and flood-control structures into one engineered unit.) Maybe this structure is quite similar to the Dutch Ooster Schelde barrier, especially the floodgates, although we will need at least one shipping lock at the highest part of the bridge over the navigation/tidal channel. Maybe it's something entirely new. To find out, put the design up for competition and wait for unfettered American genius to show its colors. This structure could have locks or gated sections to allow shipping free movement. It could address the environmental issues that helped stall the idea forty years ago, allowing normal tidal interchange to and from the lake. There is no good reason

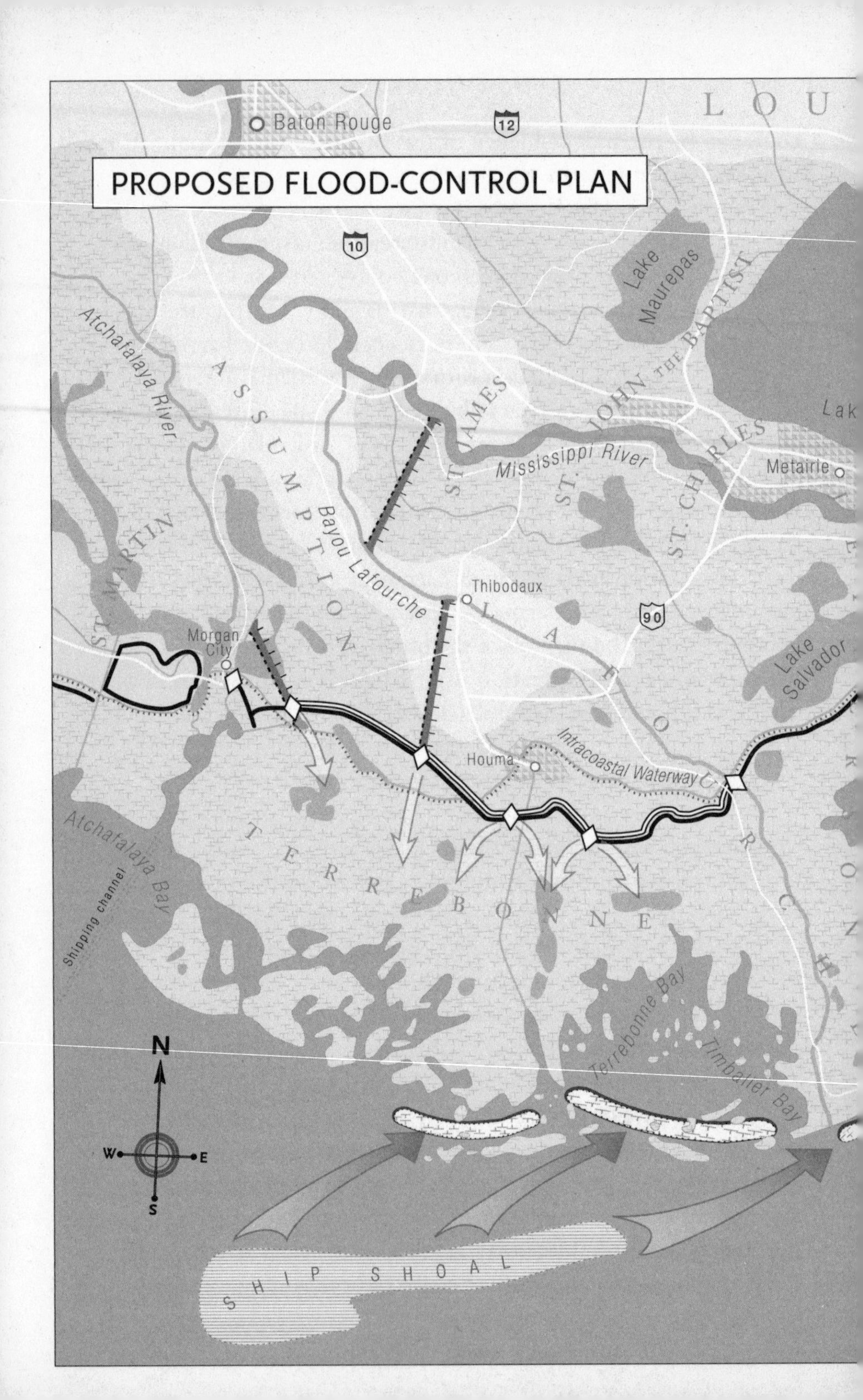

PROPOSED FLOOD-CONTROL PLAN
Baton Rouge
12
10
90
LOU
Lake Maurepas
ST. JOHN THE BAPTIST
ST. JAMES
ST. CHARLES
ASSUMPTION
ST. MARTIN
LAFOURCHE
TERREBONNE
Atchafalaya River
Mississippi River
Bayou Lafourche
Metairie
Thibodaux
Morgan City
Houma
Intracoastal Waterway
Lake Salvador
Atchafalaya Bay
Shipping channel
Terrebonne Bay
Timbalier Bay
SHIP SHOAL
N
W
E
S

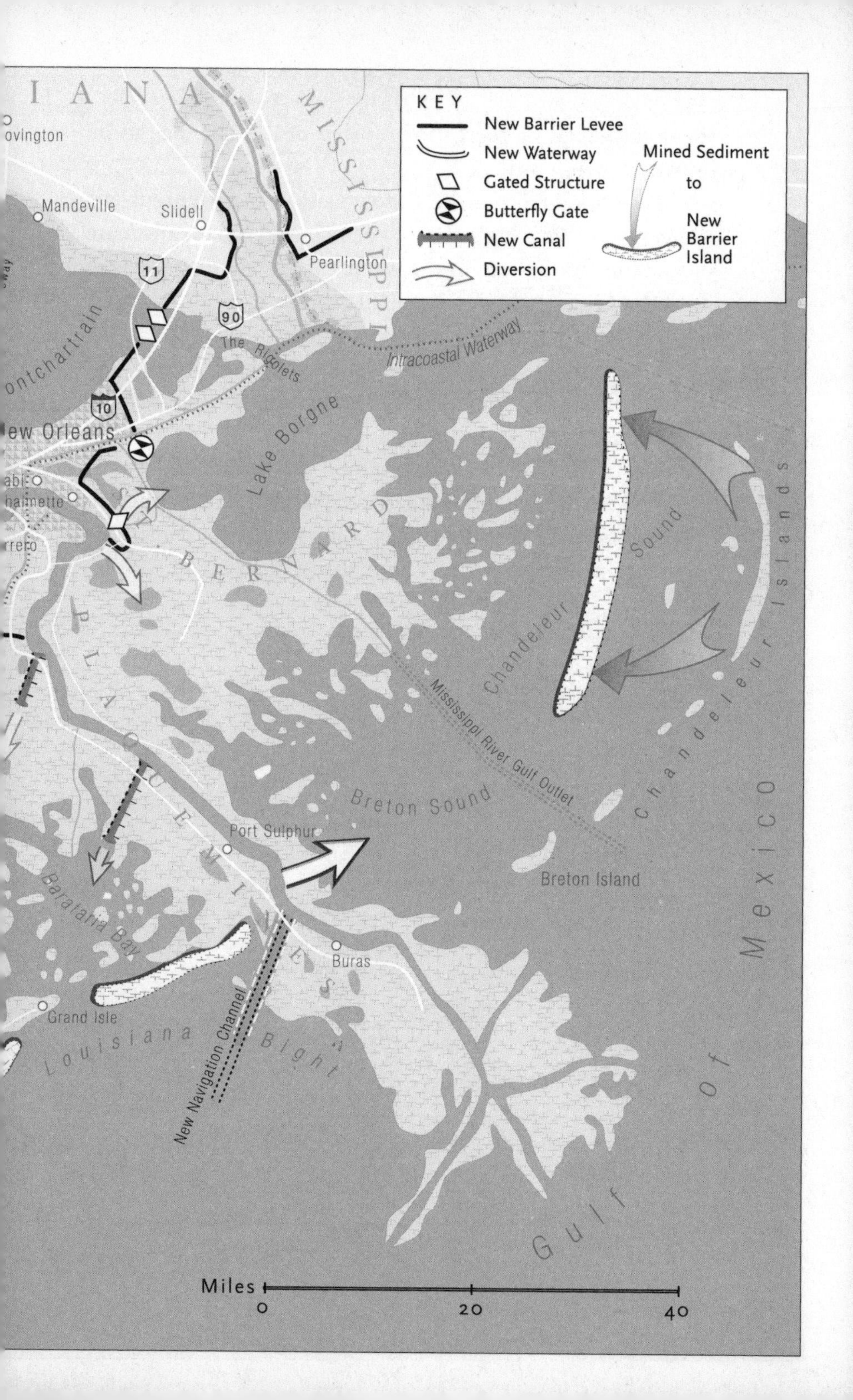

KEY
New Barrier Levee
New Waterway
Gated Structure
Butterfly Gate
New Canal
Diversion
Mined Sediment to New Barrier Island
MISSISSIPPI
Mandeville
Slidell
Pearlington
11
90
10
The Rigolets
Intracoastal Waterway
Lake Borgne
New Orleans
ST. BERNARD
PLAQUEMINES
Chandeleur Sound
Chandeleur Islands
Mississippi River Gulf Outlet
Breton Sound
Port Sulphur
Breton Island
Barataria Bay
Buras
Grand Isle
Louisiana Bight
New Navigation Channel
Gulf of Mexico
Miles
0
20
40

the commercial or recreational value of the lake has to be impaired in any way.

These floodgates at the Rigolets and Chef Menteur entrances to the lake have always been considered the ultimate protection for the city. If the water cannot get into the lake from the Gulf of Mexico in the first place, it cannot then be pushed over the levees by the north wind.

As the story is usually written, the Corps's sixties-era plan to build floodgates for Lake Pontchartrain was blocked by environmentalists. Indeed, certain green groups did file suit, asserting that the gates would inhibit tidal flows and therefore damage the fragile ecology of the lake, and an injunction was entered in December 1977, calling for a second environmental impact statement. (The first had been filed in 1974.)

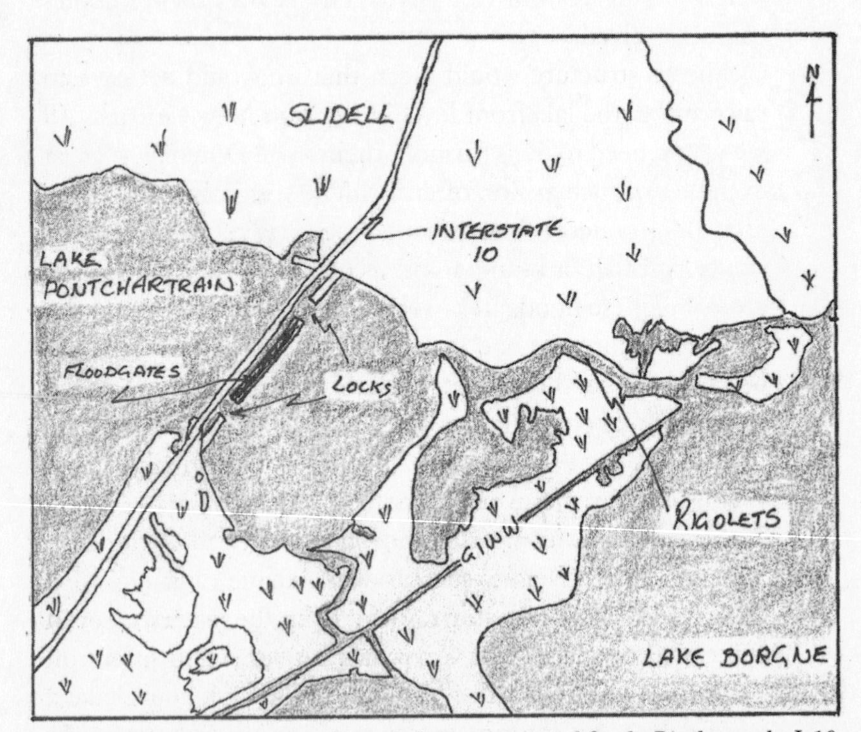

Location of surge floodgates and twin lock system proposed for the Rigolets at the I-10 Bridge. This would protect not only New Orleans but all the parishes bounding lakes Pontchartrain and Maurepas.

However, the local entities and levee boards that would have had to maintain and operate the gates were also wary. In the end, the Corps gave up the gates project and focused on the levees, which were deemed much less expensive to work with, and theoretically just as effective.

Post-Katrina, the shelved barrier plan became an issue in the campaign of various right-wing commentators to hammer environmental organizations at every opportunity. See! they screamed. If the tree-huggers hadn't stopped the barriers, this wouldn't have happened. It's not that simple. According to a statement issued by the Corps in late 2004, its engineers have been revisiting the idea of the floodgates, this time with "environmental modifications" to address the concerns that had prompted the lawsuit forty years earlier. These floodgates must be built. The environmental issues can be resolved.

Such a structure would mean that we would *not* have to raise any of the lakefront levee systems in New Orleans. All we would need to do is "armor" them. The Dutch have great examples of such armor, often a shield of tightly fitted nine-inch-thick concrete blocks with a loose gravel covering, or a layer of asphalt or bitumen mix as protection against surface erosion. These floodgates in the Rigolets would also protect the north shore of Lake Pontchartrain and communities all the way up to the Country Club of Louisiana in Baton Rouge, which could have some flooding with a Cat 5 surge. Therefore, huge areas of St. Tammany, Tangipahoa, and Livingston parishes would benefit greatly.

- The pump stations must be moved to the head of the 17th Street, London Avenue, and Orleans Avenue canals in New Orleans, blocking any storm surge from the lake from even entering the canals. (The Corps has announced its intention to do just this, at a cost of $3 billion.) The job could take a few years; as an interim measure, simple gated structures could be built. We saw several versions of such structures in the Netherlands, most featuring stout wooden gates, hinged and moved into place by hydraulic rams and locked into place with hydraulic pins. These provisional gates could be built in about a year. Obviously, lakefront gates create a problem if we have a storm with high surge and lots of rainfall. With the gates closed to protect against the surge, the existing pumps could not operate should there be flooding from rainfall, so we need new pump stations along the lakefront, fully automated, computer-driven, and with standby generators to make them immune to power failures. No one has to stay behind to operate them. The Dutch even have automatic scoop systems that lift the debris from the pump intakes and require no manual labor.

- During Katrina we saw how water poured from the Industrial Canal into Lake Pontchartrain. A gated structure with either a large lock or a surge barrier plus a lock needs to be built where the Ted Hickey Bridge crosses the Industrial Canal (Seabrook), a hundred yards or so from the lake—something similar to the Hollandse IJssel surge gate.
- Eastward from the Rigolets Barrier system the levees from Orleans East to the Intracoastal Waterway have to be substantially raised and made into major barrier levees. (Likewise, we must have a major barrier system east of the Rigolets to protect Slidell and possibly parts of coastal Mississippi.) East of New Orleans we need a butterfly-gate type of flood-control structure in the area of the Funnel. It could be a copycat version of the Dutch structure at Maeslandt (and thereby perhaps become a tourist mecca, like Maeslandt). We need substantial levees extending from this butterfly gate and linking with the great levees along the Mississippi River that protect us from river floods. It goes without saying that all levees should have armoring that saves them from both wave attack and erosion should they be overtopped.
- From the Mississippi River westward we need to build one giant barrier levee that basically follows a bulging curve seaward, taking in Houma and some of its surrounding com-

munities (some smaller curves may be necessary to incorporate some communities). With careful location, this levee, about eighty-four miles long, can balance the protection of the communities "inside" and wetlands fronting the system on the "outside." This inland levee would include numerous gated structures including, for example, where the barrier levee crosses Bayou Lafourche. Others would be built at key navigation channels and where river/sediment/freshwater diversions will be built. (See description following.) The barrier levee would then tie into the East Guide Levee of the Atchafalaya River at a point close to Morgan City. This levee can then be extended from the West Guide Levee of the Atchafalaya River all the way to Texas, with navigation gates at the required locations. Ideally, this giant levee would be on the seaward flank of a large navigation channel, parts of which would include the existing Intracoastal Waterway. This channel would allow access for heavy equipment and barges with soil, rock, and other levee-building materials.

Simply put, this levee would be the line in the sand—the real deal, a Cat 5 wonder with gates and locks where needed for navigation and/or sediment diversions for coastal restoration. To some extent this new levee would replace the misshapen network of levees now in place in this area south and west of the Mississippi River. (The existing network creates some funnel effect problems of its own. It should be replaced.) On talk radio the big idea for some time has been the ultimate levee somewhere out in the marshes. On WWL-AM, Bob and Vinny say we could start the big wall tomorrow. I don't know about that, but the final plan for flood protection in southeastern Louisiana—if it is indeed Cat 5 protection—must include some such levee.

- This whole system—barrier levee with navigation/sediment distribution gates—must be accompanied by legislation that stops development in the wetlands anywhere in the newly protected areas. Otherwise our current river flooding problems will be exacerbated, because these gates, and especially

> those at the entrance of Lake Pontchartrain, will slow down drainage of water out of the lake after major rainfall floods. I know this legislation would not be popular with certain developers, but the idea, remember, is to do it right this time. A good guide would be the principles in the paper submitted to the Pelican committee by the Association of Flood Plain Managers. Don't do anything that will worsen your neighbors' flooding. Likewise, there needs to be a moratorium on the mining of cypress forests for mulch anywhere in coastal Louisiana. Cypress swamps are some of our best defense against storm surge.

Let's acknowledge right now that many communities are going to be outside this new levee system. Some retreat has to take place. Communities outside the new system could be given some level of flood protection, but protection that costs more than the value of the protected infrastructure is, clearly, problematic, and hard to sell to the federal government, which would be paying for this job.

It would also be circumspect to require that all new construction within the new barrier be elevated a given number of feet as determined by the design specifics of the whole system. This would provide an extra margin of protection in the event of some levee overtopping, or to reduce flood damage should there be a rainfall flood that exceeds the capacity of the pumps.

The alternative to the proposed levee cutting across this part of the state is a much longer wall right at the coastline, an unbelievably expensive venture that offers no wetlands protection. I mentioned earlier that the determined citizens of Terrebonne Parish have levied a quarter-cent sales tax to raise money for their share of an ambitious, seventy-two-mile structure known as the Morganza-to-the-Gulf levee. The Corps has already spent $32 million planning this levee, but it was dreamed up without full consideration of how vulnerable the whole coast is to drowning, and without a thought given to the large-scale creation of wetlands seaward of the barrier system. The old project was deserving enough under the old vision, perhaps, but it would necessitate miles and miles of finger-like levees sticking seaward toward the Gulf. So do some of the other old ideas. As the Dutch learned, and as common sense dictates, a long set of levees set out like the fingers of your hand is a poor substitute for one solid levee across your knuckles. The latter is the much safer, less expensive system. (The old projects and studies would be extremely useful and would provide valuable data, background information, and project justification material for the major barrier system. They would allow us to fast-track the new system, because much of the ground work has been done.)

One major difference between Louisiana and the Netherlands is that we do still have a lot of wetlands for storm-surge protection. In order to protect the levee system, and as a hedge against a rising sea level and climate change, we must concentrate on building and restoring wetlands seaward of the barrier, using the sediment resources of the Atchafalaya and Mississippi rivers. Even with the major barrier system, these wetlands are still the outer defense and must be maintained for this purpose, as well as for their commercial, recreational, and environmental uses. Natural systems and some developed areas within the core should also be maintained, in order to accommodate expected overtopping and to avoid the degradation of water quality that the Dutch have experienced. Seaward of these wetlands we need to build barrier island chains, close in to the existing wetlands, using high-quality sand from offshore.

Lieutenant General Carl Strock, head of the Army Corps of Engineers in Washington, was simply wrong when he said that a restoration program would not have helped much with Katrina's storm surge because the storm passed to the east of the wetlands. Come on, General. How did the surge get into Lake Pontchartrain? How did it get into the lower part of St. Bernard Parish? By crossing thousands of acres of open water that just thirty to forty years ago had been healthy marsh and swamp. Maybe General Strock was under the spell of MR-GO, which has created a wonderful channel for the encroachment of saltwater from the Gulf of Mexico into what had been some of the most productive marshes and wetlands in the entire United States. More than eleven thousand acres of cypress swamps have been destroyed, and almost twenty thousand acres of brackish marsh have converted into less productive saline marsh. With proper management instead of desecration these regions would have diminished the surge from Katrina not only in Louisiana but in parts of western Mississippi. Levees protected by healthy marshes survive and protect people; levees exposed to open water are annihilated and leave communities open to devastation.

We need the best levees. We need extensive wetlands. And we

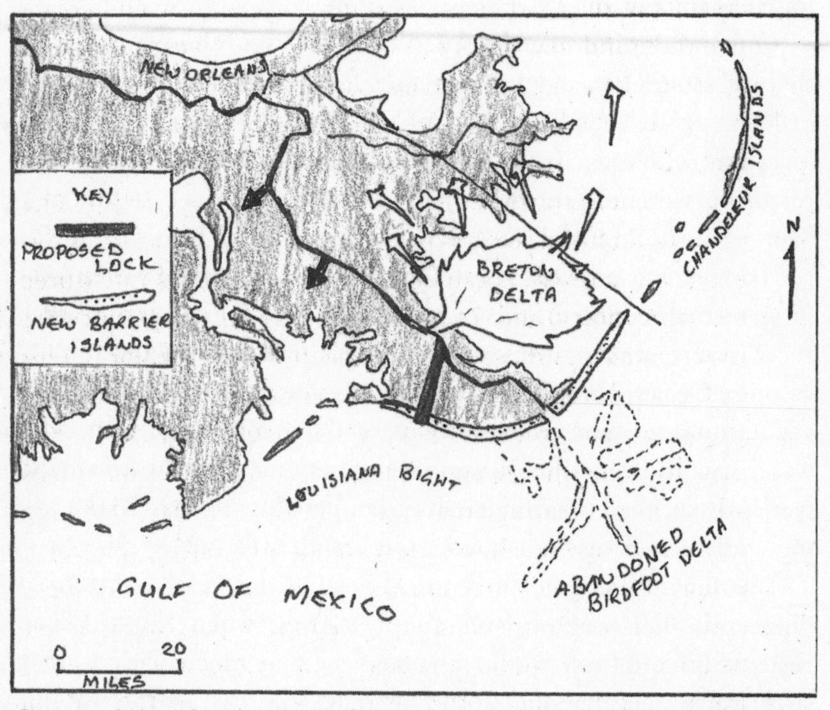

Conceptual plan of major diversions from the lower reaches of the Mississippi River, notably the Breton Delta. Also shown are the new navigation channel and the abandoned Bird-Foot Delta now reworked into a barrier island chain.

need them under one administrative roof. It would be senseless to differentiate between the two elements for either planning or financing purposes.

In Chapter 8 I mentioned the four major wetlands initiatives I set forth in the CCEER white paper in 1994. With minimal modification these are still the obvious projects we need immediately.

Here are the major elements necessary for large-scale wetland creation:

- Below New Orleans, divert the Mississippi River into the Breton and Chandeleur sounds, 32 miles upstream from the present main outlet. The Breton and Chandeleur sounds are dominated by shallow bays that have modest subsidence rates (half a foot per century) and are therefore excellent recipients of river sediment. This new diversion would create five thousand new acres of marshland every year in eastern Plaquemines and St. Bernard parishes. In twenty years we would have 140 square miles of new wetlands to buffer storm surges. In a domino effect, these healthy marshes would benefit the other marshes in St. Bernard Parish. The plan would mean the loss of the bird-foot delta, but it is dying slowly anyway, because of its high subsidence rate (three feet per century) and the absence of sedimentation, since the river discharges most of it into deep waters offshore. This dying delta could be reworked into new barrier islands that would coalesce with the Breton-Chandeleur islands to the east and the shoals and islands of western Plaquemines Parish, thus creating an almost unbroken barrier island arc south and east of New Orleans and its suburbs. The plan would also eventually require, in all likelihood, a set of locks and other control structures at the new outlet. Some oyster beds would have to be relocated.
- Reconnect Bayou Lafourche to the main channel of the Mississippi River fifteen to twenty miles to the east of its junction with the Mississippi and divert 10 percent to 12 percent of the total flow down this channel, which would feed the marshlands all along its course, bring freshwater to 8 percent of the state's population, and create jobs in the fisheries. Dredging the bayou would be necessary, along with extensive plumbing that would include small diversions and siphons—all to the immediate and lasting benefit of six hundred thousand acres of wetlands.

- Increase the flow down the Atchafalaya from 30 percent to 40 percent, while using plumbing to manage the extra water and sediment flowing into the Terrebonne, Atchafalaya, and Teche-Vermilion basins. The east-west–oriented Intracoastal Waterway is an excellent conduit to move sediment-laden water miles laterally; the boat traffic's propellers keep the sediment in suspension. The canal dredged behind the barrier levee where it did not follow the Intracoastal Waterway could act as a similar conduit. Throughout the system, large- and small-scale diversions and siphons could help mimic the natural annual floods. The system needs to be as "leaky" as possible.

Regarding these first three big-picture items, the knowledgeable, perhaps even dubious observer asks three questions immediately: Is there any history showing that such freshwater diversions actually work? Does the river still carry enough sediment to support the additional widespread dispersal? Would the new diversion into Bayou Lafourche and the increased diversion into the Atchafalaya leave enough water in the Mississippi to sustain full navigability all the way to the new outlet? The answers are yes, yes, and yes. The granddaddy of freshwater diversions in Louisiana is the Wax Lake Outlet. This canal, between the Lower Atchafalaya River and the Gulf of Mexico, was cut in 1948 as a relief valve for floodwaters that would otherwise hit Morgan City a few miles to the east. The channel was cut through wetlands and a series of small lakes with no confining levees for the last ten miles of its length. It was free to flood as it wanted. Since the diversion offered a shorter course to the Gulf than the actual river course, this channel maintained itself; that is, it required no dredging and even started to capture more water from the river. Discharge from the Wax Lake Outlet has averaged about 10 percent of the total flow of the whole Mississippi system. It should be no surprise that seaward of the outlet we have a great evolving natural delta with extremely healthy natural levee and marsh environments, a delta that expands with every flood with no help from man. The lesson here is plain and simple: Build it right and the sediment will come, and wetlands will result. Freshwater diversions have also worked for years at the Whites Ditch, LaReusitte, and West Point a la Hache siphons.

Since 1965 the Corps has been working on two other major restoration projects, using water diverted from the Mississippi and funded not by CWPPRA but under the Water Resources Development Act. They are definitely *not* Band-Aids. They are designed to stop 9 percent of projected wetlands loss, at a cost of $131 million. (By comparison, the CWPPRA projects are designed to stop 14 percent of the expected loss at a cost of $311 million.) One of the two, the Davis Pond diversion structure located on the west bank of St. Charles Parish two miles below Luling, diverts freshwater, along with dissolved nutrients and sediments, from the Mississippi River into the Barataria Basin. The objective of the project is to reduce saltwater intrusion and thereby achieve favorable salinity conditions as a means of reducing land loss in this basin. Additionally, the diversion is expected to increase commercial and recreational fish and wildlife productivity and to increase plant growth for a healthier estuarine ecosystem. To date, however, the Davis Pond diversion operates at only 10 percent of its capacity, and not all the time at that. Obviously we are not maximizing the benefits of this expensive venture. A similar project is the Caernarvon Freshwater Diversion on the boundary between St. Bernard and Plaquemines parishes southeast of New Orleans. But this project also operates at only 10 percent of its capacity and has restored a grand total of four hundred acres!

We can do better.

Regarding the second question, whether the sedimentation coming down the Mississippi would still be sufficient to support the new diversions, I defer to Richard Kesel with LSU's Department of Geography and Anthropology. In the first half of the twentieth century, Kesel judges, the river arrived at Tarbert Landing well north of Baton Rouge hauling 463 million tons of sediment annually. Thanks to the 1,300 dams that the river and its tributaries must now negotiate upstream—all of which trap sediment in the resulting reservoirs—that number has dropped by perhaps 70 percent—a huge loss, without a doubt. An amazing diminution, really. Still, the remaining sediment *is* enough to support the recommended diversions. Likewise regarding the third question. Numerous studies confirm that 45 percent of the river's present flow could be diverted and leave sufficient flow to assure navigation through the port of New Orleans and all the way down to the new outlet.

- The fourth initiative in the large-scale wetland creation philosophy is to restore and/or build new barrier islands. They're being wiped out, resulting in increased wave energy, higher storm surges, and saltwater encroachment. Much of the oil-and-gas infrastructure is old and not engineered for the resulting quasi open-ocean conditions. Exposure of this infrastructure to such forces is setting the stage for a catastrophic oil spill—just what this state does not need. The technology required to rebuild the islands has been demonstrated on Grand Isle (the only inhabited barrier island in the state), East Island (one of those in the Isles Dernieres chain), and Wine Island. Ship Shoal, a mountain of sand eleven miles south of Isles Dernieres, could be mined for the purpose at a cost of perhaps ten dollars per cubic yard.

Regarding this initiative, the objection is straightforward: Some experts don't believe the islands can be saved. They point to the recent history of the Chandeleur Islands, at the easternmost fringe of the state, twenty miles out in the Gulf from St. Bernard Parish. In 1969, these islands were *one island*, forty-five miles long. Following the battering it received that year from Hurricane Camille, most of the island was gone, leaving a string of tiny isles in its stead. By 1998, twenty-nine years later, the island had just rebuilt itself into a single whole when Georges veered east from its predicted path, undoubtedly saving New Orleans from flooding but ripping this island apart again. After Georges a $1 million grass-planting project was completed. Then came tropical storm Isidore and Hurricane Lili in

2002. The Chandeleur Islands took another hit, and then again in 2004 with Hurricane Ivan. However, after each storm the islands displayed a remarkable capacity to repair themselves, though they were a bit shorter and a little thinner. Then came Katrina, who erased them from the map.

Therefore, ask the critics of rebuilding the islands, why fight the inevitable? Barrier islands come and go. Let them. The struggle to save them is too problematic and too expensive. I disagree on both counts. We know from experience and from the storm-surge models that the islands provide excellent buffering. The barrier islands along Louisiana's south coast have, for the most part, been cut off from their sediment supply, mostly due to the navigation and oil and gas canals. Just as we refurbish beaches in Florida and along the East Coast from time to time, some also because they have been cut off from their sediment supply, so we need to and can restore Louisiana's islands. We know that a healthy barrier island with a good sand beach berm—and wide wetlands behind the berm—is a long-lived feature. We need to build the sand parts and the marshes of these islands. The marshes have been missing from the Chandeleurs. We can create the conditions to rebuild them.

Overall, rebuilding the islands would be something of a Sisyphean effort, because there would probably be losses in every big storm, but as the islands are being torn apart, they are in turn tearing down the storm surge to a measurable extent. Not restoring them dooms all remaining wetlands to a far more rapid death. And if that's not a good enough reason, consider that the islands also form the outer edges of the giant estuaries that are the basic framework supporting the very rich fisheries in Louisiana. Lose these and the fishery collapses.

To summarize: The major barrier system and large-scale wetland creation work through these initiatives, and include elements of both attack and defense:

- outer substantial barrier-levee with floodgates
- wetland and barrier island-building and maintenance outside that barrier

- maintain compartments within compartments, where such exist

To repeat, the core protected area in southeastern Louisiana would *not* have to be evacuated for a Cat 5 storm. This would not be necessary. Other areas outside this core would have a higher level of protection than exists today, but to a lesser degree than the core, and they would sometimes require evacuation. This fact can't be allowed to compromise the commitment to protect the core. So this plan is attractive because it proposes a higher level of flood protection for the principal populated areas than anyone has previously proposed. If, like the Dutch, you never have to evacuate, you can then support a level of infrastructure investment that has not been possible in the past. With this plan in place there would be no reason *not* to rebuild every single neighborhood in metropolitan New Orleans, and every reason to do so. The residents of the Lower Ninth would be as safe as those in the highest redoubts in the Garden District, which were left high and dry after the Katrina flooding. On the other hand, the plan does require political and civic courage. It requires honesty from our politicians and clearheaded, pragmatic acceptance of the facts from the citizens because restructuring the coastal landscape will mean dislocations and disruptions of some traditional fisheries, for example. Resources will be lost while new ones will be gained. Dislocations will be kept to a minimum, and the planning effort must incorporate compensation for those who lose land or livelihood. Such a fair program can assure that we will not waste years in the courts or arbitration proceedings. Am I dreaming? As I said, our citizens will have to accept the facts of life in Louisiana.

Again, this is not rocket science. The science is the easy part. The hard part is overcoming the narrow-mindedness and selfishness of politics and business as usual. For decades the two have undermined plan after plan to restore the wetlands, build new ones, and thereby protect people and property. They have played hell with improving the existing levee system. We must do better now, or we can kiss it all good-bye for good. I was not exaggerating in the introduction when I said that politics and business as usual in Louisiana will eventually put everything below Interstate 10 underwater. Science and engineering can save the day, but not if they're censored or manipulated. If that's to be the case, just shelve them and start packing. It's over.

It is February 2006. The lights are still out in the Lower Ninth Ward, Lakeview, Gentilly, Upper Ninth, elsewhere. At night these neighborhoods are dark, lonely, quiet. The diaspora from Hurricane Katrina has contracted, but only slightly. A few stalwart souls have returned and now live in FEMA trailers in their backyards, if those backyards are large enough. Jack-o'-lantern neighborhoods, these have been called, accurately. They're spooky. They may also be more dangerous than they appear. The ubiquitous layer of sludge is contaminated with arsenic and a lot else we'd be better off without. Short-term health impacts, including skin rashes, respiratory distress, and asthma, are practically a given; no one really knows the long-term impact of exposure to this mud.

Basically, most of New Orleans is still very dead. Large expanses of southeast Louisiana and coastal Mississippi still lie in ruins. Gray Line has started bus tours, and even bike tours, of the devastated neighborhoods. Is this dismayingly morbid rubbernecking or a good way to build awareness and political support for rebuilding the city (in addition to making money)? Your call. (I think they do help.)

The news about FEMA doesn't seem to get any better. Every day it's a new story about some glaring failure before or during the Katrina emergency, a new "probe" about dubious accounting, dubious contracts awarded without competitive bidding, inexplicable delays for folks still waiting for their trailers. This agency has proved that it is not capable of handling a complex catastrophe (or billions of cash dollars, apparently). Therefore, no agency in the federal government is. The military can establish a rough order—as it must—but the Katrina emergency proved that rough order is just the start. Without a radically revised attitude on the part of this and all subsequent administrations in Washington, we are doomed to repeat this debacle in southeast Louisiana, or somewhere else, or both.

What about hurricane season 2006? Just how safe is the greater

New Orleans area? The Corps's district office says it will have everything back to pre-Katrina levels by June 1. I'm not sure what this means, because the whole levee system is compromised. It will *not* be back at pre-Katrina surge-protection levels. The heights of the levees may be as they were, but height, as we saw with the catastrophic failure of the London Avenue and 17th Street levees, is not the only factor in their survival. The Corps says it will have the MR-GO levees rebuilt by June. With what? There are no sources of good material anywhere near those destroyed levee sites. Site inspections by Team Louisiana have shown that the contractors, in many areas, are using a sand base material for the repair. They appear to be scraping up the remnants of the old levees and trying to use this material, plus some they have barged in. Sand is porous and highly permeable, and it lends itself to dangerous seeps—not exactly the material of choice for levees. The contractors themselves are concerned; their field staff told us they are not getting the support or supervision they expected from the Corps. The Berkeley levee warriors and the ASCE team have been quite vocal in the media about these questionable repairs. Are they a strong pointer to the future? I sure hope not. Moreover, the progress out east has been painfully slow. Unless drastic changes are made, these levees will not be fully returned to the pre-Katrina heights, even using the unsuitable materials. There will be no armoring at all. Any wave field that develops in Lake Borgne with an east wind will start to erode these levees instantly.

At the other side of the metropolitan region, way out west at the border of Jefferson Parish and St. Charles Parish, a levee I haven't mentioned because it was not a problem during Katrina could be a terrible problem with the next storm. For a stretch of about 450 feet, the flood wall sections have sunk about 6 inches and are offset one from the next by a few inches at the top. This structure needs help, and fast. At the Industrial Canal above the Lower Ninth Ward, the I wall has been removed and will be replaced by an inverted T wall with batter piles. This is good, because it's a far more robust design. The Corps understands that the whole levee wall was compromised in this area. The three breaches on the infamous drainage canals along the south shore of Lake Pontchartrain have been sealed, but both the 17th Street and London Avenue canals now have, as a result, sections that are much narrower than before. The capacity of the canals to transport floodwater out of the city is therefore seriously diminished. I guess the Corps figures that since very few folks have moved back to the flooded neighborhoods, a bit of rainfall flooding won't hurt too much. But the real concern along these canals is that sections of the flood walls that did not breach during Katrina, but did "give" somewhat, remain in that compromised position. Early data indicates that the soils all along these levees are consistently much weaker than before Katrina, and we know that the soils contributed mightily to the Katrina collapses. So don't let them kid you. For the 2006 hurricane season, some of the levees will *not* be as strong as they were pre-Katrina.

In January 2006, New Orleans's Mayor Nagin appeared before yet another congressional committee to complain about the unending need for him to "grovel and beg" for FEMA trailers, levee money, and everything else. Then he groveled and begged one more time: "I'm puckering up! Help us!" (The mayor is on record as favoring this unique style of communication. It's refreshing, certainly, but not always helpful. I'm thinking of his subsequent avowal on Martin Luther King, Jr.'s birthday that New Orleans should always be a "chocolate" city. He soon retracted the remark and admitted a need to be more circumspect. We'll see.) The mayor's "Gang of 17" business leaders, officially the Bring Back New Orleans Commission, find themselves caught between an enormous rock and a very complex hard place. Open all devastated neighborhoods or redline some forever? How about giving former residents one year to vote with their feet, with those neighborhoods that reach some tipping point of returnees deemed viable, and all others not so? This is an audacious idea but not one likely to make the cut, or I miss my guess. It's not even clear that any government entity has the legal authority for the wholesale condemnation that would have to follow. That question is before the courts, but this hasn't stopped Sean Reilly, a member of the Louisiana Recovery Authority, from saying bluntly that city officials had no business even holding out the hope that every neighborhood would be rebuilt: "Someone has to be tough, to stand up, and to tell the truth. Every neighborhood in New Orleans will not be able to come back safe and viable. The LRA is speaking the truth with the money it controls."

On that score, though, state legislators have made it clear that they're not anxious to give up *their* historic control of the purse strings: The LRA is doing a great job, but *we* are the people's accountable representatives, and we want to stay accountable. (Strange. Politicians these days usually duck accountability. I guess it's the money here that makes them willing to take the chance.) In a special session called by the governor in November 2005, the legislators flexed their muscles by defeating a plan to dissolve the local levee boards in southeastern Louisiana and consolidate their functions under a new state authority, then they defeated a separate plan to create a commission to oversee the boards. The idea was that businesses, before they will return to New Orleans in large numbers, need confidence in a new oversight mechanism for the levees; the legislators took umbrage at the notion that their actions had anything to do with patronage and misappropriation. The Gang of 17 in New Orleans was disgusted. Two months later, in a second special legislative session, the legislators had a sudden and predictable change of heart and gave a nearly unanimous blessing to milestone legislation dissolving local boards in the New Orleans area in favor of regional authorities. The new system takes effect January 1, 2007, and is, in my opinion, a mandatory step forward in the development of the statewide big picture. Meanwhile, I'm reliably informed that the parishes have lost all trust in the Army Corps of Engineers, and in the ability of other federal or state agencies to develop and institute a comprehensive plan.

I'm on record defending the radical idea that the same federal government that drowned New Orleans with the failure of its levees should compensate all of those who lost lives and homes. Instead, the best we have come up with so far is the plan devised by Congressman Richard Baker from suburban Baton Rouge to allot homeowners and lenders 60 percent of the pre-Katrina value of their demolished property. Since the federal government is responsible for 100 percent of the losses, why not compensate for *100 percent* of them? Still, coming from a conservative Republican, this was a pretty radical, progressive idea, and is appreciated as such. Of course, the Bush administration has declared the plan dead, but in February 2006, Governor Blanco announced a state initiative that adds $4.2 billion authorized by Congress for community development block grant funding to the $6.2 billion already guaranteed. Blanco proposed a cap of $150,000 on the money available to each homeowner to repair, rebuild, relocate, or accept as a buyout. A program to register potential homeowners is currently under way.

Let's face it. A just outcome for the homeowners is highly unlikely. Lord knows there are plenty of civil cases shaping up against various defendants, including the Corps, contractors, and insurance companies, but this litigation is guaranteed to last forever, and it is unlikely to deliver full justice to those who have lost everything. The state of Louisiana cannot possibly afford to pay for the program.

It's the strange book that must end just as the story is beginning, in a way, but such is the case here, and at this early juncture I have to be counted among the pessimists. The fight for the future of New Orleans is going to be a long and difficult one. I now picture a big theme park as the end result, a plastic place without much vitality. What an incredible loss. And those with the least resources are sure to lose the most. On this score, I'm really pessimistic. And if the right decisions are not made about the levees and the wetlands, well, forget it. On this score, I'm beyond pessimistic. Consider a recent meeting I attended at the request of Lieutenant Governor Mitch Landrieu (Senator Landrieu's brother) at which I presented the big-picture plan outlined above, using the big official state map as the backdrop. Going in, I was pretty excited. Our two senators, the governor, DOTD secretary Johnny Bradberry, and others had just returned from the Netherlands, and news reports indicated that they were coming around to my ideas for adapting features of the Dutch system in Louisiana. They reportedly even saw the wisdom of cutting the Corps out of the loop for designing the system, depending on competitions instead. On matters of flood control, they certainly saw the manifest superiority of everything Dutch. I viewed my subsequent presentation as a confluence of the lessons learned from the experience of the Dutch, Hurricanes Katrina and Rita, and from twenty years of assorted wetlands restoration initiatives. (Actually, Paul Kemp, who was also at the meeting, summarized the presentation in these terms, and I agree.) But no sooner had I completed my talk than two senior members of Governor Blanco's staff, both political appointees, astounded me (and some of the others, including Paul) with comments about how the storm surge from Katrina had overwhelmed the levee design of the Army Corps of Engineers, and so on. The Corps, now and forever! *Essayons!* I could not believe what I was hearing. Had these folks talked with the boss? They then said they were in the process of developing, along with the Corps, a comprehensive plan for the state. I can tell you exactly what it will look like: the Pelican Brief all over again, in spades.

One aspect of our American way of life that is really missing in the Katrina response and recovery is our pioneering spirit that built this country, at times over almost insurmountable odds. Just three decades ago it took us to the moon. What has happened to it? Have we lost our heart and soul? Are we no longer a nation always ready to help the underdog and the less fortunate? No! The outpouring of public support during the Katrina emergency tells me that we have not lost our founding beliefs. Now we have to sustain that spirit—that pioneering spirit of old—and accept this latest challenge that nature has thrown before us, rebuilding in a sustainable way with the best of science and engineering at our disposal. My motto, one last time: The catalyst to compromise is a thorough understanding of the science.

As a nation, let's take up the "Rebuild!" battle cry. Now is the time to put politics, egos, turf wars, and profit agendas aside. We owe it to the thirteen hundred Americans who died in the Katrina tragedy. We owe it to their survivors and to all future generations. It's now or never. Let's show the world what we're all about, here in America in the twenty-first century.

AFTERWORD

In the summer of 2007—twenty-one months after Katrina hit the Louisiana and Mississippi coastlines—where do things stand in the region? The optimism and energy with which everyone initially tackled the rebuilding of their communities has been betrayed by bureaucracy and politics at almost every turn. As a direct result, energy and hope seem to be waning. Many of us now fear that the incompetence and lack of political will are approaching a tipping point from which there will be no recovery for New Orleans, specifically. I've lost track of the many recovery plans that have been put forth by this or that group in or out of government, but it's pretty clear that local, state, and national political interests, as they are currently configured, are simply not up to the task of implementing any of these utopian visions. It is impossible for me to see how we will succeed without committed leadership *from the very top*, which will not be forthcoming from this administration. Famously, President Bush's State of the Union address in January said nothing about this ongoing national tragedy. Not one word.

In the summer of 2007, the devastated Mississippi shoreline still looks devastated. A lot of the mess of shattered homes and businesses has been removed; instead we have the cement slabs of the lost structures, or the remains of pilings, all testament to the fury of Katrina's storm surge. Very few homes have been rebuilt, and a lot of the tourism structure is still missing. The casinos are back, of course, but it will be years before this formally vibrant part of the Gulf Coast really returns. In metropolitan New Orleans, large areas look almost exactly as they did right after Katrina hit. Basic services are still spotty. Only 10 to 20 percent of the homes in the destroyed neighborhoods sport FEMA trailers in their front yards, and of course it's a long, long way from that trailer to a rebuilt home. Many of those who have not returned are, by their own admission, gone for good. Others are still sitting on the fence, waiting for a serious resolution of the levee problem.

About 60 percent of the businesses have not returned. In sum, the population of the city of New Orleans has shrunk by at least 50 percent, and no one pretends that it will ever again approach the pre-Katrina level of half a million. Only 25 percent of the residents have returned to St. Bernard Parish. The only number on the way up appears to be the murder rate.

Repairs to the ruptured levees were supposed to have been completed by June 1, 2006—the beginning of that year's hurricane season. And so they were—but pathetically inadequately. The "repaired" Mississippi River-Gulf Outlet (MR-GO) levees are already eroding from mere rainfall, the back and front slopes scarred by washout rills and gullies. There is no "armoring," in many cases not even with grass (due in part to the salty nature of the soils used). In a big storm, these and sections of other levees don't stand a chance. The Corps of Engineers has installed "temporary" pumps and flood gates at the mouths of the 17th Street, London, and Orleans Avenue canals. The plan is to lower big metal gates into a system of slides when a high surge threatens. However, there have been problems in getting some of these gates to slide properly. But even assuming the gates do close properly, some of the new pumps have only 10 percent of the capacity of the very big pumps located at the head of these drainage outfall canals. Moreover, these floodwalls are now so compromised that the Corps has lowered the maximum allowable height of the water in each canal, pending a permanent fix for the floodwalls. In the meantime—years—the canals will be less efficient in emptying the big bowl of regular rainwater. So sections of the city will flood just from heavy rain; they already have in the past twenty-one months—several times. In March of 2007 it was learned that the Corps has installed defective flood control pumps in the outfall canals despite warnings from its own pump expert that the equipment would fail in a storm. The pumps were designed and built under a $26.6 million contract awarded to Moving Water Industries of Florida. One of the owners was once a business partner of former Governor Jeb Bush. (According to the Center for Responsive Politics the company has donated more than $125,000 to politicians, mostly Republicans.) In May 2006 a Corps mechanical engineer who was overseeing quality assurance at a test site in Florida sent a seventy-two-page memo stating that the equipment was defective and would break down "should they be tasked to run, under normal use, as would be required in the event of a hurricane." So this really begs the question—is New Orleans again to be doomed because of political cronyism and incompetence? Given that mere tropical storms can bring up to 40 inches of rain and gale force winds, would you feel safe in the city right now? All in all, the level of flood protection for the city of New Orleans *today*—summer 2007—is no better than it was on the day Katrina hit! A direct blow from a major hurricane will sink the city again. The main difference between then and now is that people know this and are factoring it into the decision about returning and rebuilding.

On June 1, 2006, the Corps acknowledged that the levees it had built to protect New Orleans were an incomplete and inconsistent patchwork, flawed in design and construction, never capable of handling a direct strike from a major hurricane. (As I showed in the book, Katrina was *not* that major hurricane, not in New Orleans.) A contrite Lieutenant General Carl Strock said, "This is the first time that the Corps has had to stand up and say, 'We've had a catastrophic failure.' . . . Words alone will not restore trust in the Corps." General Strock said that the agency has undergone a period of intense introspection and is "deeply saddened and enormously troubled by the suffering of so many. . . . The Corps is responsible for the projects we build and manage, and we are accountable to the American people."

These remarks were widely accepted as the confession of an honorable man who had taken responsibility and "fallen on his sword." Two months after making these remarks, General Strock resigned his post for "family and personal reasons." He was followed out the door by a number of other senior Corps officials. Of course, most if not all of the major mistakes by the Corps preceded General Strock's tenure. They date back many decades. Our research at the Hurricane Center has concluded that perhaps half of the 1,600 deaths in the Katrina tragedy are attributable to this long-term failure—that is, to the design and construction of the levees. The other half of the fatalities are more fairly attributed to the ineffective and sloppy response. This ratio is not merely an academic issue. It speaks directly to one of my major issues, as readers of this book know: the federal government's responsibility for compensation. The legalities in all this are wonderfully complex, as we would expect, but it does appear now that a constitutional "hold harmless" principle protects the Corps (and the federal government) from being sued for the deaths and destruction that resulted from their engineering mistakes in the design and construction of the flood control levees. On the other hand, legal challenges in federal court are attempting to demonstrate that the levees along MR-GO fall outside this protection, because MR-GO is primarily a *navigation* channel. These challenges won one important battle when a U.S. district judge ruled in February 2007 that the Corps can be sued for the agency's negligence in fixing the MR-GO defects it had known about for years; but there are many more to go. Other legal scholars are only now looking into the possibility of suing FEMA.

My own position—and that of many others—hasn't changed, and it is rooted in basic fairness: bottom line, the federal government was and is responsible for the levees, legal niceties be damned, and it should do everything imaginable to right the wrong, especially since such actions and infusions of money are necessary for the overall rebuilding job.

The day after President Bush ignored the Katrina tragedy in his 2007 State of the Union speech, Louisiana Governor Kathleen Blanco angrily criticized the omission and said the state is being shortchanged in federal recovery funding for political reasons. New Orleans Mayor Ray Nagin echoed her disappointment but cautioned against reopening the political rifts that became evident after the storm. He makes a good point, but probably to no avail. How do you take the politics out of anything going on in this state? How do you overlook the remarks of none other than former FEMA chief Michael "Heckuva job, Brownie" Brown, who played such a prominent role in this story? Just a week before the State of the Union speech, Brown told a group of graduate students in New York that in the earliest days following the catastrophe, some in the White House had suggested the federal government take charge in Louisiana because Blanco was a female Democrat, but stay out of Mississippi, where Governor Haley Barbour was a male Republican. Through a spokesperson, Governor Blanco reacted predictably to Brownie's remarks: "This is exactly what we were living but could not bring ourselves to believe. Karl Rove was playing politics while our people were dying. The federal effort was delayed, and now the public knows why. It's disgusting." A White House spokesperson then reacted predictably to Brownie's claims and the governor's response. Numerous disaster scientists and experts have speculated among ourselves that the failed response was due to dirty politics. Let's hope this is not true—but let's also find out. If it is true, the guilty verdict adds even more weight to the argument for just compensation for the unnecessary deaths.

Given Brownie's proven incompetence at FEMA and his understandable desire to burnish his reputation, we have the right to question the motives behind his current statements, but I am intrigued by the fact that the further we get from Katrina, the more forthcoming he seems to be about those events. I'm inclined to credit his remarks. It's also fair to say that his "allegations" are of a piece with everything else we know about how this administration in Washington operates. Nor is all this just a parlor game for the amusement of local political junkies. It has consequences on the ground. In January 2007, Governor Blanco confirmed other reports that Louisiana, burdened with 80 percent of the storm damage from Katrina and Rita, received only 55 percent of federal relief funds. FEMA had given

Mississippi, with 31,000 families living in trailers, $280 million for Katrina Cottages, while Louisiana's 64,000 families living in the trailers merited only $74 million. Clearly, I believe, politics are at the heart of the ongoing tragedy. They are sustaining it.

The federal government has appropriated $7 billion to help reimburse residents of Louisiana who have suffered major damage to their homes—apparently the largest single housing recovery program in U.S. history, as well it should be. Getting the appropriation for this "Road Home" program was the result of some very hard work by the Louisiana Recovery Authority, led by Andy Koplin, Governor Blanco's former chief of staff. Under it, eligible homeowners are supposed to receive funding to get them back into their rehabilitated dwellings. The financial award will be placed in a disbursement account, with funds to be released as related expenses are incurred. The total assistance to any one homeowner is capped at $150,000. The average payment per family so far seems to be about $40,000. Why any cap at all? No good, fair reason. Through no fault of their own, many of these families lost everything they owned: home, furniture, personal belongings, cars, and means of employment. I believe compensation should be 100 percent.

But caps aren't the main problem with the program: delivery is. ICF International, the company with a $756 million state contract to distribute the federal money, has received heavy criticism for errors and slow delivery. State legislators have recommended firing the company, and they have some astounding numbers to back up their position. Of the more than 100,000 applications received for Road Home assistance, a grand total of 359 had received grants as of late January 2007. Think about this, please: 100,000-plus applications, 359 paying accounts. There are no extra zeroes in the big number, no missing zeroes in the little one. *What is going on here?* In December, Republican Senator David Vitter said, "All I can say is Road Home . . . is a debacle, it needs to be fixed. And I have started to look, in detail, at that ICF contract and I'm startled at what's in it. The dollar amounts—$19 million travel budget . . . I don't get it."

No one does. No one gets anything about what's going on down here. (Nor is anyone surprised that the problems with the Road Home program are being used to smite our Democratic governor, perhaps with some fairness.) Funding for Road Home grants is administered separately from FEMA's public assistance payments, but the complaints about the two programs are the same. (In April 2007 it was announced that the Road Home program had run out of money, leaving tens of thousands of families without relief.) Simply put, all the mistakes and red tape that have hamstrung every aspect of the recovery effort from the beginning, as outlined in this book, persist to this very day twenty-one months later. We are no better than a banana republic. It's an unbelievable shame and embarrassment. The debacle defies rational analysis—at least it has so far.

In the book I showed how Congress turned against Louisiana when our congressional delegation succumbed to the lobbyists' greed and submitted requests for billions in cash that led to headlines around the country about "Louisiana Looters." In December 2006, it was the state politicians' turn to shoot themselves—and the rest of the state—in the foot. I speak of the one-week special session of the state legislature to spend the surprising state surplus of $1.5 billion. This bonanza was mainly attributable to sales taxes brought in by the combined Katrina and Rita recovery efforts. It was a one-time deal. The governor wanted to spend the money on pay raises, tax breaks, tax credits, and health-care issues, and she placed twenty-five issues on the agenda. After all the arguing, some of the items were not even addressed, much less appropriated. Two that were approved were $300 million for the infrastructure needs for a possible steel factory on the Mississippi River between New Orleans and Baton Rouge (It went to Alabama!), and $236 million for refunding homeowners' fees paid to the state-run Citizens Property Insurance.

Faced with a tough reelection campaign, Governor Blanco was accused of playing Father Christmas. (In March 2007 the governor announced that she would not be running for reelection.) My big question is the one I heard time and again at public forums: Since Louisiana has this coastal protection problem that must be solved, once and for all, why was none of this money earmarked for that vital purpose? Everyone at the state level is begging the federal government to own up to its mistakes and to put up the money to rectify them, but when the state itself had a one-time profit to spend, how did it proceed? Did it look to the future, or to short-term political interests? The latter.

This little special session didn't garner nearly the publicity accorded the Louisiana Looters debacle, thank goodness, but I can only assume it reinforced attitudes in Washington. It can only have raised more questions about the political ability of state and local politics to get serious about restoring the coastal wetlands as part of a comprehensive flood control system.

On the other hand, both Louisiana and Mississippi are now set to share in the revenue from the offshore oil and gas production—revenue that has previously gone entirely to the federal government. Signed into law in December 2006, the Domenici-Landrieu Gulf of Mexico Energy Security Act S. 3711 opens 8.3 million new acres in the Gulf of Mexico to oil and natural gas production and shares 37.5 percent of new revenues with the other energy-producing states on the Gulf Coast: Louisiana, Texas, Mississippi, and Alabama. The revenue sharing part of this legislation is long overdue. (That story is way too long to tell here, but the summary is simple: over fifty years ago, Louisiana politicians got greedy (imagine!), turned down President Truman's offer of 37.5 percent of the royalties, tried to get 100 percent, and ended up with none. Total royalties to date: $160 *billion*; Louisiana's share, $0.)

The new revenues are supposed to be used for wetlands restoration, hurricane protection, and flood control projects. Louisiana has tried to further buttress this dedicated commitment by passing a constitutional amendment that reinforces it. But the definition of a flood control project can be slippery, can't it? Only a fool or a coconspirator would leave the selection to a bunch of lobbyists.

Under the new law, Louisiana is projected to receive at least $13 billion over the next thirty years, but most of the money won't come for at least a decade. The state is expected to receive a few billion dollars for hurricane protection, wetlands restoration, and navigation projects in the next ten years from the regular budget process and other previous legislation, such as the Breaux Act, the Coastal Impact Assistance Program Senator Landrieu championed in 2005 (approximately $135 million to Louisiana per annum), as well as the Katrina- and Rita-related supplemental appropriations bills which Senator Landrieu helped to craft from her position on the Senate Appropriations Committee.

This offshore revenue sharing is major money, without a doubt, and all to the good at first glance, but the Domenici-Landrieu Act, specifically, was opposed by environmental groups arguing that conservation should not be tied to expanded offshore oil and gas drilling, given the repercussions of such drilling. Lawsuits could be filed; the arguments do have some merit. Inevitably, the drilling will add to the pollution of Gulf Coast shorelines and habitat and feed our ultimately unsustainable appetite for hydrocarbons and the global warming consequences of their burning. And why should coastal restoration be tied to oil and gas exploration? In the ideal world, it wouldn't be. In the world of Louisiana politics—which are dominated by oil and gas interests, obviously—it's probably inevitable. In any event, it's what we have now. The task before us is to make sure that this money actually goes to restoration, that it is joined by the rest of the total required, that a master plan is drawn up and authorized, and that the work begins.

Is this too much to ask of the system down here? We can only hope that the result of the Katrina recovery effort so far is not the answer.

Lieutenant General Carl Strock's apologetic remarks about the Corps's levees were offered on the occasion of the agency's release of its 6,000-page report on the disaster (officially known as the Interagency Performance Evaluation Taskforce—IPET). Despite General Strock's statement, a key finding of this IPET report is that the catastrophic failure of the levee and floodwall section at the 17th Street Canal was in part the result of two *unforeseen* issues: (1) the presence of an unusually weak layer of clays at the toe of the levee, and (2) the fact that the floodwall was pushed inboard (away from the canal) by the rising waters, creating a gap between the floodwall (including the steel-sheet pile curtain upon which it was supported) and the adjacent levee embankment soils. Water rushed into this gap, applied additional pressure against the lower sheet piles, and pushed the entire structure sideways.

At the press briefing accompanying the release of the report, Corps engineers claimed that neither of these unforeseen mechanisms had previously been reported in the scientific and engineering literature. This claim proved to be absolute, disingenuous nonsense! The independent National Science Foundation (colloquially referred to in the book as the Berkeley team) jumped on the excuse and pointed out succinctly, "[T]he U.S. Army Corps of Engineers had a masterful knowledge and understanding of the complex and challenging geology of this region in the 1950s, and they should have recognized the presence of weak clay layers."

Regarding the second claim, the "unforeseen" nature of the mode of failure, well, that's equally ludicrous. In 1985, the Corps *itself* had performed a very expensive, full-scale field test of a 200-foot long test section of a levee and sheet pile-supported floodwall. The whole idea of this ambitious test was to develop insight for the subsequent design of these absolutely vital levee and floodwall sections in the years that followed. (I did not know about this test when writing the book. Almost no outsiders did.) The embankment and the sheet pile-supported concrete floodwall were sized and built to model conditions similar to those expected along the 17th Street, Orleans, and London Avenue canals, as well as along major portions of the Inner Harbor Navigational Channel. A cofferdam scheme allowed the water level to be progressively raised on the outboard ("water") side of the floodwall. The foundation soils consisted of normally consolidated soft, highly plastic clays, conditions well-chosen to represent those expected to be encountered at the New Orleans levees themselves.

In the tests, when the water approached eight feet there was a rapid and dramatic increase in the deflections at the heads of the sheet piles as the water tried to push the piles over. A gap developed between the piles and the soils, allowing water to flow in and exert additional hydrostatic pressures on the piles. Failure was imminent. This was *exactly* the mode of failure that the Corps claimed twenty years later had been "unforeseen." Experimental data from this 1985 test were subsequently analyzed extensively by the Corps of Engineers Waterways Experiment Station, and multiple Corps of Engineers reports were issued that resulted in significant recommended revisions to the analytical methods and processes used to analyze both the soils supporting the sheet piling and concrete floodwalls, and the sheet pile/floodwalls themselves. Results from these studies were later even more widely reported in two papers in the *Electronic Journal of Geotechnical Engineering* in 1997. The first of these papers succinctly observed, "As the water level rises, the increased loading may produce separation of the soil from the pile on the flooded side (a 'crack' develops). Water pushed into the crack produces additional hydrostatic pressures on the wall side of the crack and equal and opposite pressures on the soil side of the crack."

How, then, could the Corps claim so many years later that the sheet pile failure mode had not previously been seen or published, and that it could not have been anticipated?

To this day, the Corps still claims that overtopping of the levees was a critical factor in the failures of many of them. Wrong, said the NSF/Berkeley team report issued in late July 2006. Speaking about the findings, Berkeley professor Ray Seed said, "These levees were not overtopped, they failed, primarily as a result of human error. The hurricane wasn't much bigger than the levees were designed for."

Neither the 6,000-page IPET nor any other Corps report has ever acceded to the clearly established relationship between the Katrina flooding and MR-GO. Ignoring a wealth of undisputed real-world data from hurricanes Betsy, Camille, and Katrina, ignoring the latest and best computer simulations, the Corps relies primarily on simulations based on incomplete, hypothetical scenarios. The Corps does propose closing MR-GO but has failed to recommend any hurricane protection measures for its most dangerous segment, the infamous Funnel.

Bob Bea, also of Berkeley and coauthor of that group's report, said, "One of our major findings is that the levee system protecting New Orleans was defective as a result of dysfunctional organizations. The Corps needs to be modernized, and federal and state oversight of flood control restructured, because they can't build a safe levee with current processes." His group's report plainly states that the Corps did not adequately oversee the entire levee system and laid off many of its geotechnical engineers who could have effectively fulfilled this responsibility. Instead, levees were planned and built without adequate oversight.

The Corps's IPET report tried to deny responsibility by stating in the executive summary that there was no evidence of malfeasance or negligence! But how can a purely scientific and engineering report conclude there is no evidence of wrongdoing? It can't, as the National Academy of Engineering/National Research Council Committee pointed out in its Third Report on the IPET activities, released in October 2006: "portions of the Executive Summary reach premature closure on important topics, gloss over uncertainties and differences of opinion, and assume a tone that is at times legal in nature."

Why did the Corps even bring up the issue of "negligence" and "malfeasance"? That's easy. Both are terms are fraught with ramifications in the courts: the federal government is trying to cover its backside.

Yet another group whose investigation played a significant role in the story and this book was the one working under the auspices of the American Society of Civil Engineers (ASCE). Issued two months after the Berkeley team's report, the ASCE investigation also bluntly criticized the Corps's self-serving self-exoneration. Regarding the Corps's claim that "[t]he designs for these structures were marginal with respect to practice," the ASCE group said this characterization is "far too generous in that it minimizes information documented by IPET suggesting that serious design mistakes were made." The ASCE panel "believes that some designs were much too close to the margins of safety and that protection of the public safety and welfare was not always the top priority when some choices and decisions were made."

And then there was Team Louisiana, the official state investigation into the levee failures. I was a member of this team, and our work figured prominently in my narrative. We released our final assessment in March 2007. Why so much later than the other groups? I'll tell you why: money. Everyone else had it, we didn't. Moreover, despite e-mails from senior Corps officials promising full cooperation, most of the really important documents we needed were never released to us (including the one hundred boxes of internal Corps documents that we knew about but could not gain access to [and which the Corps team did have access to]). Therefore we decided to wait for the Corps to release its own report, from which we hoped to glean at least some of the missing material. And we did. Another reason to wait was to see if the Corps would pick up some of the "bait" we had dropped along the way. For instance, the stopped-clock data, which could be used to establish the times of levee breaches and the rapidity of flooding, as I've described. As we had hoped, the Corps got into the clock data collection with a vengeance, so we could mine their report for the data we did not have the resources to collect on our own. Finally, as the small fellows on the block, we knew the bigger lads would pounce on anything we got even halfway wrong, so we did not want to be rushed to conclusions.

The wait was worth it. Among other important findings, Team Louisiana was able to show that the New Orleans Office of the Corps of Engineers:

- Used obsolete (1959) hurricane surge design parameters, even though the head office in Washington, D.C. had ordered them to use more up-to-date data in 1981; as a consequence, their surge estimates were at least 40 percent too low.
- Admitted to the General Accounting Office (now the Government Accountability Office) that they knew their surge estimates were too low, but still stayed right on course.
- Misinterpreted surveyor elevation datum, therefore started building the levees 1.5 feet too low! Then after realizing their mistake in 1985, they *continued* to build them 1.5 feet too low.
- Totally ignored their own design manuals in terms of protecting levees from wave erosion, that is, no armoring of the levees, especially those made of highly erodable sand and shell.
- Failed to undertake rudimentary seepage calculations, especially along the London Avenue Canal, which is underlain by highly permeable sand. Where seepage calculations were made, they were based on a technique known as the Lane's Weighted Creep Ratio, which all the design manuals of the time stated should not be used in the design of structures intended to protect human lives.

Simply put, New Orleans was a catastrophe waiting to happen. That's now indisputable. All the levee study groups have recommended that there must be outside, impartial, transparent review of all Corps projects—repair and upgrading of existing levees, building new levees and other structures, *everything*. This is a sad—tragic—state of affairs, but what's the choice, really?

One absolutely vital requirement for the Katrina recovery—but one that is glaringly absent from the dialogue—is the plan for true Cat 5 protection for the region. Lip service is all we have in the works now.

In its August 2006 Call to Action—the report that castigated the Corps's work in the past—the ASCE report reiterated what I and others have been yelling as loudly as we could for many years: the serious deficiencies in the southeast Louisiana hurricane protection system must be corrected if the region is to avoid a similar or even worse catastrophe than Katrina in the foreseeable future. The report states that there are flaws in the way the system was conceived, budgeted, funded, designed, constructed, managed, and operated. That pretty much covers it. Overcoming these deficiencies will require leadership, courage, conviction, and funding. I, for one, was gratified to see this group come down so powerfully on this issue. Obviously, it is close to my heart and expertise.

The Corps of Engineers was given $8 million to develop a first draft of a comprehensive plan by June 2006. A document did appear by that date, but to call it a "plan" of any meaningful sort would be a wild exaggeration. It's nothing but words crafted, I'm willing to bet, by political operatives in the White House Office of Management and Budget. That's what it reads like. Meanwhile, the state of Louisiana's own restoration and flood control team appears to have been selected by political cronyism. They released their conclusions in November 2006. A joint report by the National Wildlife Federation and the Environmental Defense (NWF/ED) hammered the work for lacking "a compelling vision of a restored coast and restoration processes."

I concur. So much for restoration vision. What about the protection side of the state's "plan"? In essence, it consists of two lines of levees cutting across the state from east to west, one providing protection from the five-hundred-year storm (that is, the incredibly damaging storm that could be expected once every five hundred years) and the one-hundred-year storm. This two-tiered solution is a policy prescription of simply tremendous import, of course, yet no rationale for it is provided. None at all. The NWF/ED report states, "We have yet to see serious analysis showing that such a large and complex system of levees is either affordable or technically feasible. There is simply no evidence that the plan is the outcome of a systematic scientific and engineering assessment of what wetland restoration and hurricane protection measures can work effectively together at a cost that is feasible and on a time line that will be genuinely useful."

Specifically, the issue of population retreat is not considered in the state plan. People have been retreating from the lower Louisiana coast for decades; after hurricanes Katrina and Rita, this retreat necessarily picked up pace, and continued retreat of some sort should be considered as a given for the future. Politicians may not understand this, but the people who vote for them already do. The details on the retreat affect the degree of hurricane protection that is required; the exact parameters of the retreat might also allow more aggressive coastal restoration approaches. Any plan that is too squeamish to get serious about this is, well, not a serious one.

On the other hand, the state's plan does list the completion of the Industrial Canal lock as a desirable project. This lock is an economic development project, and a questionable one at that. Most experts consider it a boondoggle. In any event, it is not a coastal restoration or flood protection project and has absolutely no place in any plan addressing that purpose. Its presence here says a lot about the forces maneuvering behind the scenes in this state initiative. The head of the planning effort is a former official of the Corps of Engineers. (There is a bit of good news here. Reorganization of the formerly Byzantine levee board system replaced a multitude of fiefdoms with just two, which are moving forward with some efficiency.)

This state plan is truly out of the blue, primarily because it was developed (or concocted, more accurately) without broad input from the people who know the issues. I've already noted in the last chapter that the state's own Team Louisiana was basically banned from the process. Not to toot my own horn, but to date the only comprehensive surge protection plan on the table—one that encompasses both restoration and new engineering—is the one briefly sketched in the last chapter of this book, and about which I have heard no serious rebuttal. That's not to say it's perfect, but it is an honest effort to look *only* at the science, the technology, and the economics. Lobbyists played no part in its development. This is what we need: independent expertise. As I've said, none of this is rocket science. The *politics* may be, but not the work itself. Protecting Louisiana from any conceivable future hurricane is very doable.

Non-profit organizations, scientists, scholars, present and former public officials, and concerned citizens recognized in early 2007 that there is a very real need for an independent planning effort to recommend an action plan for a comprehensive hurricane protection system for Greater New Orleans. They are talking and meeting as I write; their ideas will soon gel in a set of recommendations—with priorities—that will take advantage of the dozens of studies, reports, and critiques of the inadequacies of the current hurricane protection system. Hopefully, these will shortly be presented to Congress as a template for action.

The recommended plan should utilize state-of-the art design and technology, and be subjected to extensive expert review and public comment. This is what we need. Wish lists from the usual suspects with ulterior motives is not! (Nor do we need to be shifting already appropriated flood control money from one job to another, which President Bush proposed in March 2007, to the tune of $1.3 billion. About this unbelievably shortsighted scheme Senator Vitter complained, "I am deathly afraid that this vital emergency post-Katrina work is now being treated like a typical [Corps] project that takes decades to complete. We will not recover if this happens.")

We have had a 9–11 Commission. How about an 8–29 Commission to fully explore just how we got to the situation where the levees that once (or were supposed to) offered protection from the three-hundred-year storm, according to the federal government, couldn't hold back a ten-year storm? The issues raised in the various forensics studies could form the basis for this official inquiry, but it would differ substantially from any of the technical investigations that have been concluded. The forensics reports do not tell us *why* decisions were made in the way they were and therefore cannot build confidence in the Corps of Engineers, which will have a lot to do with whatever restoration protection is implemented.

Do we really need yet another investigation? Yes, if it will get to the root causes of the political and bureaucratic *disease* that has brought us to this tragic, pivotal juncture. Yes, if it will be followed by committed presidential leadership. (Maybe we can get this by 2009. We shall see.) Levees.org, one of the most influential citizens' groups to emerge from the Katrina detritus, started a petition drive in support of an 8–29 Commission early in January 2007. Coincidentally, I and others had already called for a Katrina Bill that will fully compensate those who lost everything, as well as local governments, businesses, hospitals, universities, and so on to allow them to get reestablished. Coastal restoration should be fully federally funded, as should coastal protection. Nether should be tied to the potential revenues from oil and gas royalties. Additionally, we suggest a reexamination and careful restriction of the unique blanket immunity that the federal government and its Corps of Engineers enjoys from liability for defective design of flood control projects.

Finally—and of least importance, lord knows—a few words about my whistle-blowing problems with LSU, which hit the national news briefly in 2006. Right after the Corps of Engineers issued its self-exonerating report in June, LSU Vice Chancellor Michael Ruffner sent letters to several major newspapers claiming I had "no professional credentials or training" to discuss the engineering of levees. This was in response to a May 29 article in the *New York Times* written by John Schwartz titled "Ivor van Heerden's *Storm* Draws Fire at L.S.U." Schwartz was one of the reporters placed off-limits to me by university officials because my criticism of the Corps, FEMA, and other agencies in the months following Katrina was apparently hurting the school's ability to raise federal dollars. So much for academic freedom at Louisiana's flagship university. (I noted this incident in the book.) *Times*man Schwartz had used a freedom of information request to obtain forty-three e-mails from LSU covering the episode. Subsequently he discussed the matter with Vice Chancellor Ruffner and Chancellor Sean O'Keefe. In their meeting with Schwartz, he told me, neither LSU official disputed a single fact in his article.

Forty-three LSU professors wrote the local media in defense of my free speech rights *and* my expertise on the questions at hand. They said, "[T]his faculty member is listed as an associate professor in LSU's civil and environmental engineering department and as deputy director of the Hurricane Center. One would think such credentials were satisfactory enough to speak on levees and hurricanes—as van Heerden had done long before Katrina. Furthermore, detailed studies have now vindicated his criticisms.

"The attempt to muzzle a professor seems to have been motivated by worries that criticism of powerful men and agencies may jeopardize federal funding to LSU. This anxiety loses sight of what is really important, namely that human error and incompetence caused a good part of Hurricane Katrina's terrible death and destruction, and that recognizing past mistakes is necessary to avoid future ones.

"Universities have a special mission, even a duty, to examine and speak on all matters, especially on those that affect the public. Academic freedom means that faculty members are free to investigate and discuss issues . . . administrative attempts to determine the outcome of research by intimidating professors tarnish the institution."

Other newspapers wrote editorials about the whole affair, none friendly to LSU. To my mind, the crowning comment came from the Baton Rouge online magazine 225BatonRouge.com toward the end of 2006, with a short piece written by Jeff Roedel under the byline "The Year in Blunders—Faculty Support Taken to a New Level." The article stated, "Worried about securing future federal funding for the university, LSU Vice Chancellor Michael Ruffner decides the *New York Times* opinion page is the best place to trash Team Louisiana spokesman and LSU Hurricane Center deputy director Ivor van Heerden. After all, van Heerden's an outspoken critic of post-Katrina red tape. Van Heerden's analytical dissection of Katrina, *The Storm*, has since been hailed as a revelation by scientists and the *Times* itself."

This was all very gratifying but failed to rectify the fact I have lost my right to teach; likewise, public outreach and administering big projects are no longer part of my job description at the LSU Hurricane Center. My contract has been restricted to one-year renewals, down from three, and my state salary support could be taken away at any moment. When this paperback edition is published for the second anniversary of the Katrina tragedy, I have no idea where, or even if, I will be working in my chosen field. In a closely related development, Paul Kemp has resigned from LSU (while retaining his involvement in Louisiana as a vice president of the National Audubon Society, Gulf Coast region). The Hurricane Center is being decimated, but LSU has its eye on the real prize: when the billions in restoration monies eventually arrive, as much as 10 percent could go for research. Who will hand out these research bucks? Probably the Corps of Engineers! *Essayons!*

LSU Chancellor Sean O'Keefe, a great friend of President Bush's, heads up a panel tasked by Congress to set the Corps's priorities for the future. Two LSU vice chancellors are also providing input for the panel, which is organized through the National Academy of Public Administration. Moreover, it appears that the federal government is moving toward establishing a very large and well-funded hurricane research program. LSU will be in line to garner a major chunk of this change. Could the school be determined to close down the existing hurricane center, with its nasty independence from the Corps, in order to improve its chances at scoring the really big bucks in the offing? Could there be related significance in the announcement in December 2006 that the university will reorganize its coastal research groups into a new Coastal Systems and Society Initiative under the leadership of one of the two vice chancellors mentioned above? I think so. For one thing, none of the affected or relevant LSU scientists were asked to participate in drawing up this new initiative.

All in all, I'd say the LSU administrators are winning—but from their narrow perspective only. They dream that their school will become a top-tier university, but the actions of some of them in the van Heerden episode may well provide the "karmic fodder" that will doom LSU to tier 3 status forever, as a senior administrator from a famous West Coast university suggested to me.

We lucked out with the 2006 hurricane season—barely a stiff breeze made it into the Gulf. We may luck out in 2007... 2008 ... 2015 ... 2025. But maybe not, and one of these years our luck will definitely let us down, and in order to be ready for that change of fortune with Mother Nature we need a real protection plan and the serious money to make it happen (serious money, but still a pittance compared with what we are spending in Iraq). Moreover, the system-wide failure that was Katrina will happen again somewhere else unless system-wide changes are made—with the Corps, with FEMA, with every pertinent agency. The Democrats took control of Congress in January 2007. I have been urging citizens groups to work directly with the new leadership right now, because they're all we have in Washington, D.C. If a competent president—Democrat or Republican—with a vision of the importance of New Orleans takes office in January 2009, so much the better. One point is indisputable: If the leaders of the future also fail us, you can forget this part of the United States of America. It's gone.

ACKNOWLEDGMENTS

I want to thank my family for everything they've done for me. Katrina changed many lives and certainly ours. My wife Lorie gave me all her love and support and read early drafts of the book; daughter Julia contributed many ideas about content and cover; and younger Vanessa helped in collating copies and in other ways. Since Katrina I have missed many family events and I thank my family for their understanding. Late one night, Lorie told me she would make up for me when the book pulled me out of our home. That was the kind of encouragement I needed.

I would not have been in the position to write this book if it was not for wonderful friends such as Paul Kemp and Marc Levitan. Paul has always been a great comrade-in-arms, and thanks to Marc's vision we got into the hurricane business. They both also offered valuable suggestions on early drafts of this book.

Hurricanes Katrina and Rita really taxed our research teams, both during the response phase and now in the recovery phase. Our twin hurricane centers' researchers, all extremely dedicated, developed a thorough understanding of the impact of a major storm's hitting New Orleans, and their research was crucial throughout all aspects of this catastrophe. All gave hours and hours of time and many still do. These include Brian Wolshon and Chester Wilmot (*nog 'u suid afrikaner*)—transport and evacuation; Jeanne Hurlbert and Jack Beggs—sociology and public opinion surveys; John Pardue and Bill Moe—water contamination; Erno Sajo—air contamination; Nan Walker, Larry Rouse, and Ric Haag—Earth Scan Lab; Kevin Robbins, Barry Keim, Luigi Romolo, and Jay Grymes—Southern Regional Climate Center; John Snead, DeWitt Braud, Hampton Peele, Lisa Pond, and Robert Paulsell—GIS experts and mapmakers; Paul Kemp, Hassan Mashriqui, Joannes Westerink (Notre Dame), and Rick Luettich (UNC)—ADCIRC surge modeling; Joe Suhayda—surge modeling; Dane

Dartez—storm-surge measurements and field trip organizer; John Pine—hazard assessment and risks; Martin Hugh-Jones—veterinarian; Jim Diaz—medical doctor and tropical diseases; and Bruce Sharkey and Elizabeth English—architecture. Graduate students Ezra Boyd, Young Souk Yang, and Carol Hill really helped out as well. LSU vice provost Chuck Wilson offered early encouragement and Kristine Calogne and Michelle Spielman did sterling work under trying conditions.

I would be remiss if I did not acknowledge the Hurricane Public Health Centers research associates, Kate Streva and Ahmet Binselam. Both these folks offer so much support and are always willing to go the extra mile. I will never forget their efforts during the first few days in getting the GIS mapping going. They helped to make the HEF center the success it is and I thank them for all they still do to keep things on track.

Brian Ropers-Huilman, William Scullin, Sam White, and Steve Brandt, Center for Computational Technology, really helped move the ADCIRC modeling forward with smooth access to LSU's supercomputer.

During Katrina there were huge demands for GIS data analysis and mapmaking. Fantastic GIS support came from Barrett Kennedy, College of Art & Design; John Anderson, Andrew Curtis, and Farrell Jones, Geography and Anthropology, as well as a number of their graduate students.

I acknowledge Pace Laboratories' Dean McInnis and Sonny Macaisz, who ensured our water sampling in the first hectic days went so smoothly. Sergeant Bill Thomas of the LSU police got us into and out of New Orleans safely.

All of this research would not have been possible without the support of the Louisiana Board of Regents (BOR)—Health Excellence Fund to whom we are extremely thankful, as to Jim Gershey, BOR project manger, who gave us the go-ahead day one to use our funds for response activities. A sincere thanks to all at the BOR.

Thanks to Tom Berg of Hurricane Alley for spaghetti plots on page 68.

My trip to the Netherlands was funded by WWL-TV out of New Orleans and I thank their news manager, Mark Swinney; newsman and anchor, Dennis Woltering; and all the other reporters and support staff there for making this extremely important trip possible. On the ground in the Netherlands, Professor Jurjen Battjes of the Delft University of Technology organized the whole trip, and I am also extremely indebted to Professor Han Vrijling, Bas Jongman, students Elwyn Klaver and Wim Kanning, and cameraman Ivo Coolen.

Louisiana state officials Terry Ryder, Johnny Bradberry, Ed Preau, Susan Severance, Burton Guidry, Will Crawford, Mike Stack, Justin Guilbeau,

Amber Thomas, and Giuseppe Miserendino helped move the levee studies forward. I am indebted to the members of Team Louisiana who were willing to step into the fray: Paul Kemp, Wes Shrum, Radhey Sharma, Hassan Mashriqui, Ahmet Binselam, Billy Prochaska, Louis Capozzoli, and Art Theis. I and the state are truly indebted to the University of Berkeley surge warriors, especially Bob Bea and Ray Seed. Governor Kathleen Babineaux Blanco always had kind words of support.

The idea to write a book came as I was putting together an e-mail to friends and family late one night shortly after Katrina struck, an e-mail written partly as therapy to deal with my own feeling of sadness and partly to inform people about what was really happening in New Orleans. John Barry encouraged me to start a book and introduced me to Wendy Wolf at Viking, who, sight unseen, felt the potential in my Katrina story and ultimately brought Mike Bryan to my front door. I thank John for his support and wise counsel, and I also thank CNN's Anderson Cooper, who, during a depressing helicopter flight over New Orleans a week into the event, encouraged me to write. Wendy needs a special word of thanks, not just as my editor but also as someone who grew up in New Orleans and has a real love for the city. Wendy has been so, so supportive and encouraging as well as making sure we met deadlines.

I knew I would not have enough time to write the whole story myself and needed someone who would feel the passion of the story, who could convey that as well as the sadness. Tall, lanky, inquisitive Mike Bryan stepped into our lives on a moment's notice late September 2005. Within three and a half months we had written a book together and he had become part of the family. Mike is a wonderful teacher and friend. Mike lost his dad while we were working on the book and yet still managed to keep his eye on the ball. I am eternally indebted to "Mr. Mike," as my kids affectionately call him.

At Viking I wish to acknowledge the heroic work that was done in production, art, and design by Tricia Conley, Nancy Resnick, Carrie Ryan, Herb Thornby, Clifford Corcoran, and Rachel Burd. Adrian Kitzinger worked quickly to give us the crucial maps for the book. Attorney Jack Weiss, a Louisiana native, helped me understand how things looked from different points of view. My own attorney, Donald Price, reviewed the LSU passages and I thank him for that.

My agent, Joe Spieler, got the passion from the get-go. I thank him for his savvy advice and for sorting out the paperwork—not one of my strong points.

The media did a wonderful job of getting the word out and still does. They were a great source of data during the first few months of our levee assessments. Thanks from all of us in Louisiana.

Last, on a very personal note, my family and I are great believers in prayer, and it is through prayer that I could face some of what I saw and dealt with during the Katrina event. My prayer now is that this book does make a difference.

Ivor van Heerden
Satsuma, Louisiana
March 2006

REFERENCES AND RESOURCES

From my perspective, the hundreds of newspaper, magazine, and television reporters and producers dispatched to Louisiana and Mississippi during the Katrina emergency did a wonderful job and provided a vital public service. Their reports from the scenes are one of the three main sources for the information in this book. I've sometimes cited certain specific stories in the text, but I am not footnoting the hundreds of others because the same facts and quotes appeared in so many different stories. (For the curious, there's always Google.)

The other two main sources of information are my personal experience during the emergency and my background knowledge acquired over the years. My experiences as presented here speak for themselves; of course, this also holds true for my general knowledge of the situation along the coastline. For all errors of fact, blame me and me alone.

Below is a brief—very brief—selection of sources for further reading and research.

BOOKS

Barry, John M. *Rising Tide: The Great Mississippi Flood of 1927 and How It Changed America*. New York: Touchstone, 1997.

CNN News. *Katrina: State of Emergency*. Kansas City: Andrews McNeal Publishing, 2005.

Colten, Craig E. *An Unnatural Metropolis: Wresting New Orleans from Nature*. Baton Rouge: LSU Press, 2005.

Emanuel, Kerry. *Divine Wind: The History and Science of Hurricanes*. New York: Oxford University Press, 2005.

Larsen, Erik. *Isaac's Storm: A Man, a Time, and the Deadliest Hurricane in History*. New York: Vintage Books, 2000.

Nicholson, William C. *Emergency Response and Emergency Management Law.* Springfield, IL: C. C. Thomas, 2003.

Sheets, Bob, and Jack Williams. *Hurricane Watch: Forecasting the Deadliest Storms on Earth*. New York: Vintage Books, 2001.

Van de Ven, G. P., ed. *Man-Made Lowlands: History of the Management and Land Reclamation in the Netherlands.* Utrecht: International Commission on Irrigation and Drainage, 2004.

WEB PAGES

America's Wetlands: http://www.americaswetland.com/

ADCIRC Development Group: http://www.nd.edu/nadcirc/index.htm

CCEER 1994 Plan:
http://publichealth.hurricane.lsu.edu/Adobe%20files%20for%20webpage/CCEER%201994.pdf

Center for the Study of Public Health Impacts of Hurricanes:
http://www.publichealth.hurricane.lsu.edu

Coalition to Restore Coastal Louisiana: http://www.crcl.org/

Hurricane Alley: http://www.hurricanealley.net/

Hurricane Basics: http://hurricanes.noaa.gov/prepare/

Louisiana Environmental Action Network: http://www.leanweb.org/

Louisiana Geological Survey: http://www.lgs.lsu.edu/

Louisiana Water Resources Research Institute: http://www.lwrri.lsu.edu/

LSU Earth Scan Lab: http://www.esl.lsu.edu/home/

LSU Hurricane Center: http://www.hurricane.lsu.edu/

LSU Hurricane Experts: http://www.lsu.edu/pa/mediacenter/tipsheets/hurricane.html

National Hurrricane Center: http://www.nhc.noaa.gov/

NHC Tropical Analysis and Forecast Branch Radiofax Broadcast Schedule: http://www.nhc.noaa.gov/radiofax.shtml

Southern Regional Climate Center: http://www.srcc.lsu.edu/

"Washing Away" Series, *Times Picayune:* http://www.nola.com/hurricane/?/washingaway/

Weather Underground: http://www.wunderground.com/

INDEX

Page numbers in *italics* refer to illustrations